SCIENCE WORKSHOP SERIES

Biology

Revised Edition

HUMAN BIOLOGY

Seymour Rosen

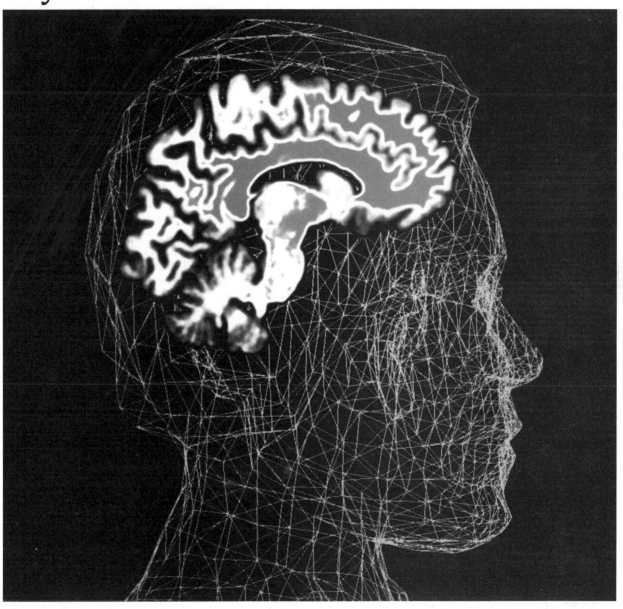

GLOBE FEARON

Pearson Learning Group

W0010631

THE AUTHOR

Seymour Rosen received his B.A. and M.S. degrees from Brooklyn College. He taught science in the New York City School System for twenty-seven years. Mr. Rosen was also a contributing participant in a teacher-training program for the development of science curriculum for the New York City Board of Education.

Cover Designer: Joan Jacobus
Cover Photograph: Alfred Pasieka/Photo Researchers, Inc.
Cover Photo Researcher: Joan Jacobus
Photo Researchers: Rhoda Sidney, Jenifer Hixson

About the cover illustration: This brain image, generated by a heat-sensitive device, is paired with a computer-generated image of a human head. There are three main parts of the brain. The cerebrum is primarily responsible for conscious thought and memory. The cerebellum is primarily responsible for balance, posture, and coordination. The brain stem controls actions such as breathing and reflexes.

ISBN 0-130-23381-1

Printed in the United States of America

3 4 5 6 7 8 9 10   05 04 03 02

Globe Fearon
Pearson Learning Group

1-800-321-3106
www.pearsonlearning.com

CONTENTS

Introduction to Human Biology

Can you think of a machine that burns fuel for heat and energy and has such a strong pump that it works for years and years without stopping? A car? An engine? No! It is your own body!

A human being can do many things, from running a marathon to dreaming about space travel. But some of the most amazing things happen inside your body. This busy "machine" can fight against invading germs. It can change complicated food chemicals into simple ones. It sends important materials from one part to another. It sends messages from one part to another, too. And most of these things are happening while you are sleeping, working, or watching TV!

In this book you will learn about how to plan balanced meals. You also will learn how the body deals with foreign substances, such as drugs, alcohol, and tobacco.

However, most importantly, you will learn how decisions you make today, will affect your body in the future.

What are tissues and organs?

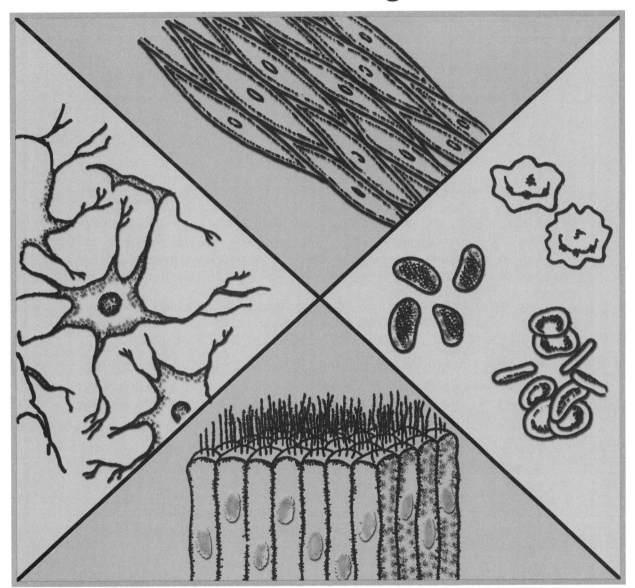

KEY TERMS

specialized cells: cells that are similar in size and shape

tissues: group of similar cells that work together to perform a specific function

organs: groups of tissues that join together to do a specific job

LESSON 1 | What are tissues and organs?

A car has many parts. Each part does a special job. All the parts must work together to keep the car running well.

In some ways, your body is like a car. Your body has many parts. These parts work together to keep you running well.

As you know, your body is made up of trillions of cells. These cells are alike in certain ways. But the cells are not all the same. They have different sizes and shapes. Different kinds of cells have different jobs. They are **specialized** [SPESH-uh-lyzed] **cells**. Specialized cells are similar in size and shape. The shapes of most cells help them to do their jobs.

The job of a specialized cell can be done only by that kind of cell. No other kind of cell can do that job. For example, only nerve cells can send and receive messages. Only muscle cells can make bones move.

TISSUES In many-celled organisms, cells work as teams, just like players on a baseball team. They form specialized groups of cells called **tissues**. A tissue is a group of similar cells that work together to perform a specific function.

Humans are made of four main kinds of tissues. They include epithelial [ep-uh-THEEL-ee-uhl] tissue, nerve tissue, connective tissue, and muscle tissue.

ORGANS Groups of cells that work together form tissues. Different tissues also "team up." Groups of tissues that join together to do a specific job are called **organs** [OWR-gunz].

Your body has many organs. Your heart is an organ. It pumps blood throughout your body. The heart is an organ of circulation. Your nose, windpipe, and lungs are organs, too. These organs are used for respiration. You also have sense organs. Sense organs tell you "what's happening," both inside your body and outside your body.

Use what you have read so far to answer the questions below.

1. What combine to form tissues? _____

2. Name four kinds of tissue found in the human body. _____,

   _____, _____, and _____.

3. What combine to form organs? _____

HUMAN TISSUES AND THEIR SPECIAL JOBS

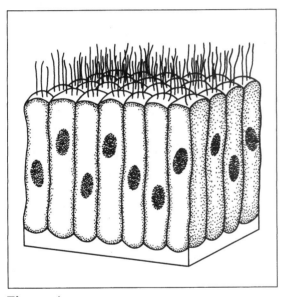

Figure A

EPITHELIAL TISSUE is covering tissue. It is made up of cells that are joined tightly together. Skin is made of epithelial tissue. Epithelial tissue covers organs both inside and outside your body. It helps keep out germs, and protects you from injury.

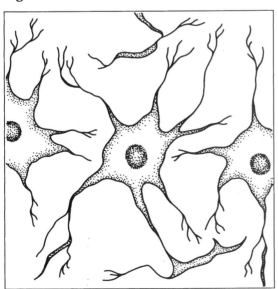

Figure B

NERVE TISSUE is made up of nerve cells. It sends and receives messages. Nerve tissue allows us to respond to <u>stimuli</u> [STIM-yuh-ly], or changes in our <u>surroundings</u>. Nerve tissue responds to changes both inside and outside the body.

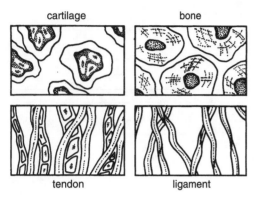

Figure C

CONNECTIVE TISSUE supports the body and holds it together. Connective tissue also helps to protect the body.

Bone, cartilage [KART-ul-idj], tendons, and ligaments [LIG-uh-ments], are all examples of connective tissue.

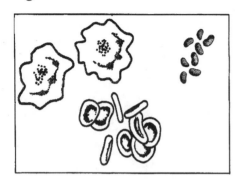

Figure D

Blood also is a connective tissue. It is a liquid connective tissue. Blood carries oxygen, digested food, and important chemicals to all parts of the body. Blood tissue also carries away wastes.

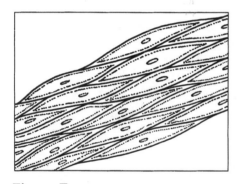

Figure E

MUSCLE TISSUE makes movement possible. Muscle tissue is made up of cells that can become shorter. There are different kinds of muscle tissue. One kind is attached to bones. When these muscles shorten, they pull on bones.

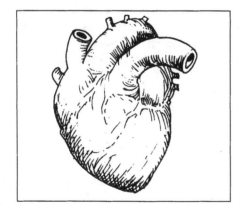

Figure F

Your body has many organs. An organ is made mostly of just one kind of tissue. But an organ has other tissues, also. For example, your HEART is an organ. It pumps blood throughout your body. The heart is made mostly of muscle tissue. But it is also made of blood tissue, nerve tissue, and epithelial tissue.

The chart below lists several organs and their jobs. It also lists some of the tissues that make up each organ.

ORGANS	JOB	TISSUES
HEART	pumps blood through-out the body	mostly muscle; also blood, nerve, and epithelial
STOMACH	digests food	muscle, nerve blood, and other tissues
SKIN	covers and protects the body; helps get rid of salts, water, heat, and a small amount of urea; prevents loss of body fluids	mostly epithelial; also blood, nerve, and other tissues
BRAIN and SPINAL CORD	the brain is the organ of thinking; the brain and spinal cord send and receive messages	mostly nerve; also blood, connective, and other tissues
EARS, EYES, NOSE, TONGUE, and SKIN	sense organs; tell what is happening around you	nerve, muscle, blood, and other tissues

WHAT DO THE PICTURES SHOW?

Use the information in the chart above to answer the questions about the pictures below.

The two organs shown here are made mostly of nerve tissue.

1. What is the name of organ A?

   _____

2. What is the name of organ B?

   _____

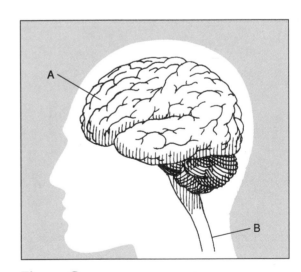

Figure G

5

Your skin is the largest organ in your body. Sweat glands in the skin get rid of waste products.

3. What are three jobs of the skin?

_____

_____

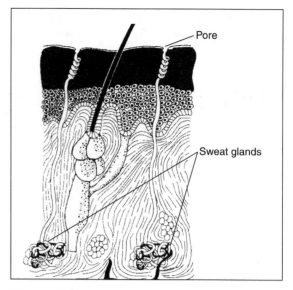

Figure H

MATCHING

Match each term in Column A with its description in Column B. Write the correct letter in the space provided.

	Column A		Column B
_____	1. blood	**a)**	pumps blood
_____	2. connective tissue	**b)**	covering tissue
_____	3. epithelial tissues	**c)**	made mostly of nerve tissue
_____	4. stimuli	**d)**	carries oxygen and food to cells
_____	5. sense organs	**e)**	produces movement
_____	6. muscle tissue	**f)**	bone, tendons, ligaments, and cartilage
_____	7. brain and spinal cord	**g)**	organ of digestion
_____	8. stomach	**h)**	ears, eyes, nose, skin, and tongue
_____	9. lungs	**i)**	organs of respiration
_____	10. heart	**j)**	changes in our surroundings

What is an organ system?

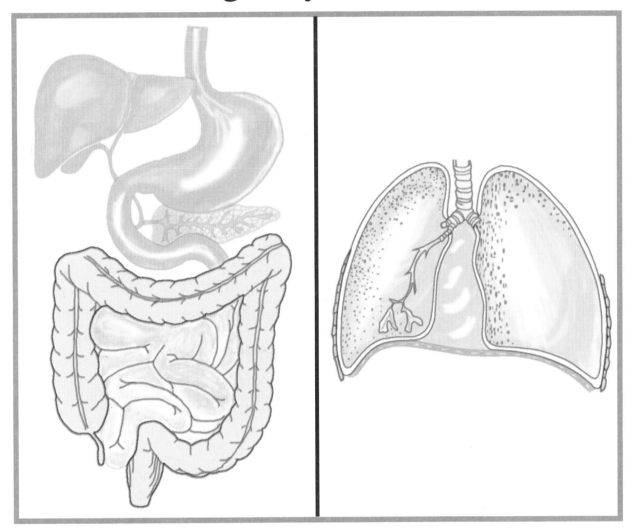

KEY TERMS

organ system: group of organs that work together

digestive system: body system that changes food to a form the body can use

excretory system: body system responsible for removing cellular wastes from the body

respiratory system: breathing system that brings oxygen into the body and gets rid of carbon dioxide

LESSON 2 | What is an organ system?

As you know, cells do not work alone. Similar cells join together to form tissues. Tissues, in turn, form organs. Organs do important jobs. But even organs do not work alone.

Usually, several organs work together to carry out a specific life function. A group of organs that work together to do a specific job is called an **organ system**.

Each organ in an organ system has a specific job to do. For example, the mouth, the esophagus, the stomach, and the small intestine are organs of the **digestive** [di-JES-tiv] **system**. The digestive system changes food to a form the body can use.

The human body has several other organ systems. Each organ system works to carry out one of the life functions.

Look at the chart on the next page. As you look at this chart, you will notice that some organs are part of more than one organ system.

For example:

- The liver is part of the digestive system. The liver is also part of the **excretory** [EKS-kruh-towr-ee] **system**. The excretory system gets rid of cell wastes.

- The mouth is part of the **respiratory** [RES-pur-uh-towr-ee] **system**. The respiratory (breathing) system brings oxygen into the body. It also gets rid of carbon dioxide. As you already know, the mouth is part of the digestive system, too.

Your body has several organ systems. These systems all work together. And, together, they form a living organism—YOU!

ORGAN SYSTEMS AND THEIR ORGANS

ORGAN SYSTEM	MAJOR ORGANS
Digestive system	mouth esophagus stomach small intestine large intestine liver pancreas
Respiratory system	nose and mouth trachea lungs (2)
Circulatory system	heart blood vessels
Nervous system	brain spinal cord
Excretory system	kidneys (2) skin lungs liver large intestine bladder
Reproductive system	ovaries (2) (female) testes (2) (male)
Endocrine system	thyroid gland pituitary gland thymus
Muscular system	muscles
Skeletal system	bone

The pictures below show what you have just read about: organs, organisms, and organ systems. Which is which? Write the correct label on the line under each picture.

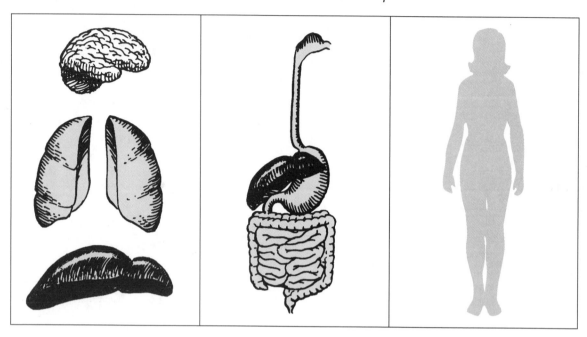

Figure A

Figure B

Figure C

_____   _____   _____

GLANDS

The organs shown in Figure D are **glands**. Glands make up the endocrine system. Glands make chemicals the body needs to carry out the life functions.

1. What organ system do the organs in Figure D belong to?

   _____

2. What organs of this organ system does a female have that a male does

   not have? _____

3. What organs of this system does a male have that a female does not

   have? _____

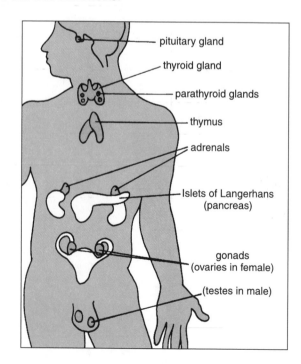

pituitary gland

thyroid gland

parathyroid glands

thymus

adrenals

Islets of Langerhans
(pancreas)

gonads
(ovaries in female)

(testes in male)

Figure D

The organs of some of the organ systems are listed in the first column of the chart. Make a check (✔) in any of the other columns if the organ belongs to any of the systems. Remember: An organ may belong to more than one system.

	ORGANS	ORGAN SYSTEMS							
		diges-tive	respira-tory	excre-tory	repro-ductive	circula-tory	nervous	endo-crine	skeletal
1.	large intestine								
2.	bladder								
3.	brain								
4.	ovaries								
5.	nose								
6.	liver								
7.	blood vessels								
8.	kidneys								
9.	spinal cord								
10.	lungs								
11.	heart								
12.	small intestine								
13.	mouth								
14.	bone								
15.	trachea								
16.	esophagus								
17.	skin								
18.	testes								
19.	stomach								
20.	thyroid								

FILL IN THE BLANK

Complete each statement using a term or terms from the list below. Write your answers in the spaces provided. Some words may be used more than once.

respiratory   tissues   excretory
organ system circulatory   digestive
organ   more than one organism

1. Cells group together to form _____ .

2. Tissues working together make up an _____ .

3. Two or more organs working together make up an _____ .

4. Organ systems make up an _____ .

5. Some organs work in _____ system.

6. The liver is part of the _____ system. It is also part of the

   _____ system.

7. The large intestine is part of the _____ system. It is also part of the

   _____ system.

8. The heart is part of the _____ system.

9. The lungs are part of the _____ system. The lungs are also part of the

   _____ system.

MATCHING

Match each term in Column A with its description in Column B. Write the correct letter in the space provided.

	Column A		Column B
_____	1. kidney	**a)**	organ of the nervous system
_____	2. ovaries	**b)**	organ of the excretory system
_____	3. spinal cord	**c)**	group of organs working together
_____	4. organ system	**d)**	part of digestive and respiratory systems
_____	5. mouth	**e)**	organs of the reproductive system

What is the skeletal system?

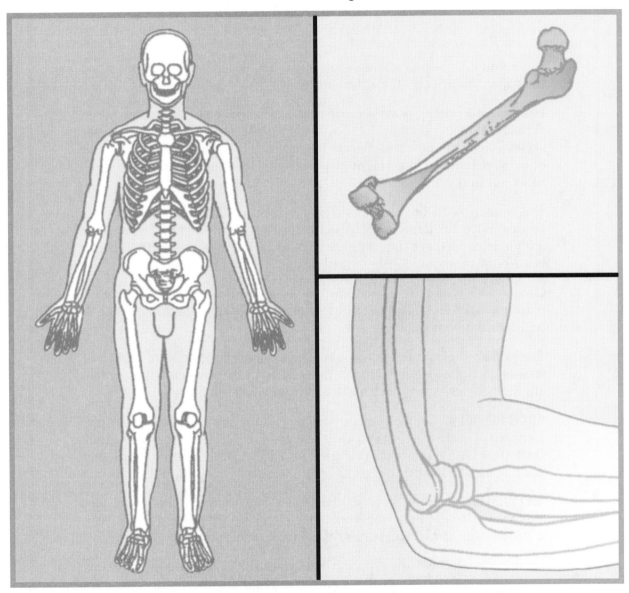

KEY TERMS

cartilage: tough, flexible connective tissue

joint: place where two or more bones meet

ligaments: tissue that connects bone to bone

marrow: soft tissue in a bone that makes blood cells

LESSON 3 | What is the skeletal system?

Have you ever seen a house being built? The first thing that goes up is the frame. It supports the entire house.

Humans, and many other animals, have a frame, too. This frame is the skeleton. Some animals, like crabs and insects, have a hard **outer** skeleton called an **exoskeleton** [ek-soh-SKEL-uh-tun]. Humans, and other vertebrates, have an **internal** skeleton, or **endoskeleton** [en-duh-SKEL-uh-tun].

The human skeleton is made mostly of bone. It also has some softer tissues called **cartilage** [KART-ul-idj]. Your ears and the tip of your nose are made of cartilage. Squeeze them gently. They can move. You cannot bend bone that way!

Cartilage also lines the inner surface of most **joints**. A joint is the meeting place of two bones. Cartilage in the joints acts like a shock absorber. It cushions the bones.

The human skeleton has 206 bones. The skeleton supports the body, but it does even more. For example, the skeleton also protects vital organs, allows free movement, and makes red and white blood cells.

PROTECTION   Think about your body. Your brain, heart, and lungs are three of your vital organs. These organs are protected by bones. Your skull protects your brain. Your ribs and breastbone (sternum) protect your heart and lungs.

MOVEMENT   Some joints are moveable. Other joints are not moveable. For example, the joints of your skull are not moveable. The joints of your arms, legs, hands, and feet, however, are moveable.

Most joints are held together by **ligaments** [LIG-uh-ments]. Ligaments stretch easily. This allows the bones to move easily. Bones and muscles work together to produce movement.

BLOOD CELL PRODUCTION   Some bones have tubelike canals. They are filled with soft tissue called **marrow**. Red blood cells and some white blood cells are made in the bone marrow.

Figure A

Figure A shows many of the 206 bones of the human skeleton. Study the diagram. Then answer the questions.

1. **a)** The human skeleton is an

   _____ skeleton.
   internal, external

   b) What do we call an internal

   skeleton? _____

2. The human skeleton is made mostly

   of hard _____ tissue.

3. **a)** What do we call the flexible tissue that makes up some parts of the skeleton?

   _____

   b) Name two parts of the skeleton that are made of this tissue. _____

   _____

4. Look at Figure A again. Find each of the bones listed below. Then write the scientific name for each of these bones.

   a) kneecap _____

   b) shin bone _____

   c) skull _____

   d) breastbone _____

   e) jaw bone _____

   f) hip bone _____

   g) collarbone _____

   h) shoulder blade _____

   i) backbone _____

   j) thigh bone _____

5. What two bones make up the lower leg? _____

6. What is the name of the place where two bones meet? _____

7. Which bone is most important for talking? _____

8. What bones make up the spinal column? _____

9. Identify each of these joints. Write the letter of the joint on the line next to its description.

   a) knee joint _____   e) ankle _____

   b) elbow _____   f) jaw joint _____

   c) wrist _____   g) hip joint _____

   d) shoulder joint _____

10. Number 9 points to cartilage.

    a) Which bones does this cartilage connect? _____

    b) Why must these parts be made of cartilage? _____

    _____

11. Part 10 of the skeleton are also cartilage.

    a) Which bones do these cartilage "disks" connect? _____

    b) Why are these cartilage disks important? _____

    _____

JOINTS

Bones move only at joints. There are three main kinds of joints in the body. They are fixed joints, partly-moveable joints, and moveable joints. Fixed joints do not allow any movement. The joints of your skull are not moveable. Partly-moveable joints allow a little movement. The joints between your ribs move a little. However, most of the joints in the body are moveable joints. There are four kinds of moveable joints. They are described below.

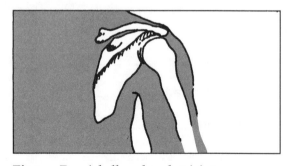

Figure B *A ball-and-socket joint*

A ball-and-socket joint can be twisted. It permits movement in many directions. This includes rotating movements. The shoulder joint is an example of a ball-and-socket joint.

1. Name another ball-and-socket joint of your body. _____

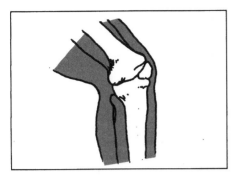

Figure C *A hinge joint*

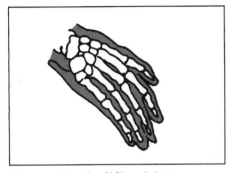

Figure D *A gliding joint*

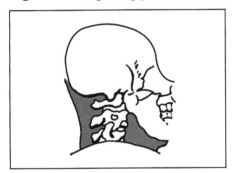

Figure E *A pivotal joint*

A <u>hinge</u> <u>joint</u> can move in only one direction, like a door hinge. The knee is an example of a hinge joint.

Bend your elbow.

2. How many directions can your elbow

bend? _____

3. Name another hinge joint in your

body. _____

A <u>gliding</u> <u>joint</u> allows some movement in all directions. Your wrist has gliding joints.

<u>Pivotal</u> <u>joints</u> allow bones to move side-to-side and up-and-down. The joint between your skull and neck is a pivotal joint.

MATCHING

Match each term in Column A with its description in Column B. Write the correct letter in the space provided.

Column A

_____ **1.** backbone

_____ **2.** shoulder joint

_____ **3.** elbow joint

_____ **4.** cartilage

_____ **5.** marrow

Column B

a) hinge joint

b) connects moveable bones

c) made up of vertebrae

d) fills some bone canals

e) ball-and-socket joint

FILL IN THE BLANK

Complete each statement using a term or terms from the list below. Write your answers in the spaces provided.

joint	skull	blood cells
bones	spinal cord	ball-and-socket
outer ears	cartilage	breastbone
movement	ligaments	protect
hinge	internal	nose
support	ribs	

1. The human skeleton is an _____ skeleton.

2. The human skeleton is made up of 206 _____ and some

    _____ .

3. The _____ and the tip of the _____ are made of cartilage.

4. Bones serve four purposes. Bones _____ , _____ , allow

    _____ , and make _____ .

5. The brain is protected by the bones of the _____ .

6. The heart and lungs are protected by the _____ and

    _____ .

7. The backbone encloses and protects the _____ .

8. The place where two bones meet is called a _____ .

9. Two kinds of moveable joints are the _____ joint and the

    _____ joint.

10. The bones at moveable joints are connected to one another by _____ .

REACHING OUT RESEARCH

Not all white blood cells are made in the bone marrow. There are two other parts of the body that make white blood cells. Use an encyclopedia, or another resource book, to find out what other parts of the body make white blood cells. (Hint: White blood cells are also called leukocytes [LOO-koh-syts].)

_____

_____

What is the muscular system?

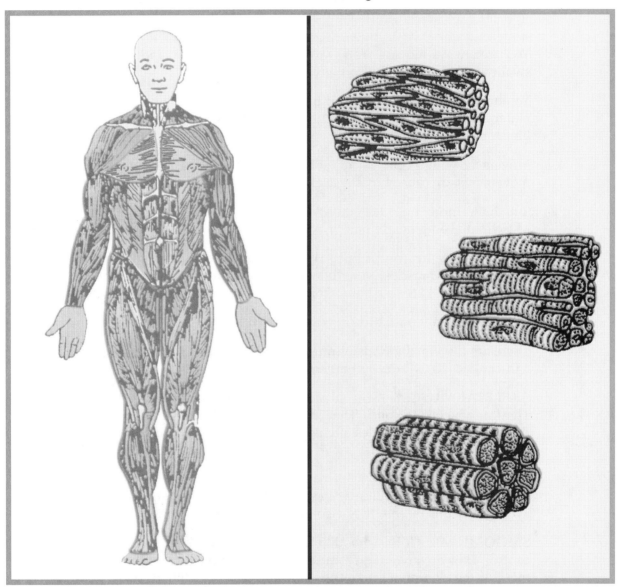

KEY TERMS

skeletal muscle: muscle attached to the skeleton, making movement possible

smooth muscle: muscle that causes movements that you cannot control

cardiac muscle: type of muscle found only in the heart

LESSON 4 | What is the muscular system?

You are always moving. You walk, you talk, you write, and chew. You swallow and blink, and also wink.

You breathe day and night. Your heart is always beating. Every moment, materials within your cells are "on the go." Movement stops only when life stops.

Movement within cells is caused by chemical reactions. All other body movements are caused by muscles.

You have more than 600 muscles. Some are large. Some are small. Muscles make up nearly half of your body weight.

Muscles work by contracting. When a muscle contracts, it shortens. Without your muscles, your bones could not move. When a muscle contracts, it PULLS on a bone. This pulling action produces movement.

Muscles can only pull. Muscles cannot push.

There are three main kinds of muscle: **skeletal muscle**, **smooth muscle**, and **cardiac** [KAHR-dee-ak] **muscle**.

SKELETAL MUSCLE   Skeletal muscles are muscles that you *can* control. They are attached to bones. They move when you *want* them to move. For this reason, skeletal muscles often are called voluntary [VAHL-un-ter-ee] muscles. The muscles that move your arms and legs are examples of skeletal muscles.

Under a microscope, voluntary muscles look striped or striated [STRY-ayt-ed]. For this reason, they also are called striated muscles.

SMOOTH MUSCLE   Smooth muscles are muscles that you *cannot* control. They are involuntary muscles. Smooth muscles form the walls of most of the digestive tract. They also are found in blood vessels and other internal organs. How do you think smooth muscles look under a microscope?

CARDIAC MUSCLE   Cardiac muscle is heart muscle. Under a microscope, cardiac muscle appears striated like voluntary muscles. But, cardiac muscle is involuntary. You have no control over cardiac muscle.

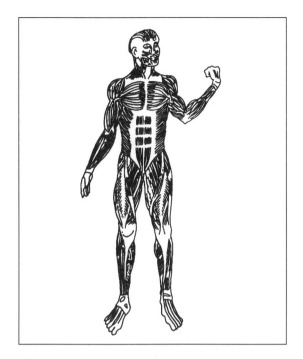

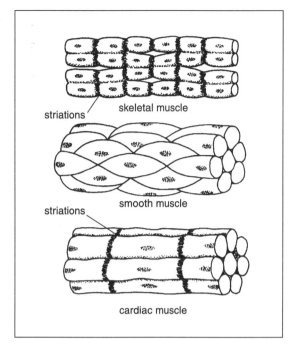

Figure A   **Figure B**

Answer these questions about the human muscular system.

1. About how many muscles does a person have? _____

2. Name the three main types of muscle. _____, _____,

    _____

3. Muscles that we control are called _____ muscles.

4. Muscles we cannot control are called _____ muscles.

5. "Cardiac" means _____ .

6. Cardiac muscle is _____ .
       voluntary, involuntary

7. Which kind of muscles are attached to bones? _____ .
                                               voluntary, involuntary

8. Under a microscope,

    a) voluntary muscles and cardiac muscle look _____ .
                                                   smooth, striated

    b) muscles forming the walls of the digestive tract look _____ .
                                                             smooth, striated

9. What do muscles produce? _____

10. Muscles produce movement by _____ .
                               pushing, pulling

Skeletal muscles work in pairs. One muscle straightens a bone at a joint. The other muscle bends the joint.

For example:

• The contracted muscle PULLS. This causes movement.

• While the contracted muscle pulls, the other muscle RELAXES.

It must! Otherwise, there would be no movement.

Let us work with two actual examples. Study Figures C and D. Then answer the questions.

Figure C shows some of the muscles of the arm. The muscles that bend and straighten the arm are good examples of muscles working together.

1. Name the paired muscles that bend

   the elbow. _____

   _____

Now, think carefully about these questions.

2. To bend the arm,

   a) which muscle must contract?

   _____

   b) which muscle must relax?

   _____

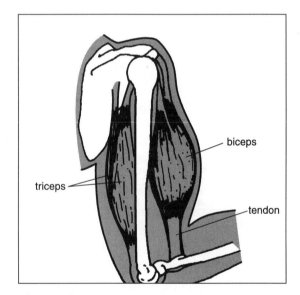

Figure C

3. To straighten the arm,

   a) which muscle must contract?

   _____

   b) which muscle must relax?

   _____

4. Most skeletal muscles are attached to bones by means of special "go-between"

   tissue. That is very strong. What is the name of this tissue? _____

Figure D shows some of the muscles of the leg.

5. Write the common names of the muscles that move the ankle.

    _____, _____

6. To bend the ankle,

    a) which muscle must contract?

    _____

    b) which muscle must relax?

    _____

7. To straighten the ankle,

    a) which muscle must contract?

    _____

    b) which muscle must relax? _____

8. What kind of muscle is the calf muscle? _____
    skeletal, smooth, cardiac

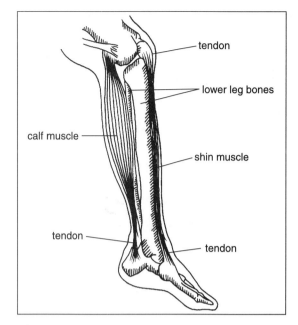

Figure D

SOME HEALTHY ADVICE

Muscles must be used often to be kept healthy. Regular, moderate exercise—and a good diet—help keep all your muscles in good shape. This includes your heart.

Figure E

Complete each statement using a term or terms from the list below. Write your answers in the spaces provided.

cardiac   bones   push
voluntary   pull   smooth
striated   muscles   involuntary
relaxes   contracts   pairs
tendons

1. Body movements are caused by _____ .

2. Muscles that you can control are called _____ muscles.

3. Muscles you cannot control are called _____ muscles.

4. Voluntary muscles are connected to _____ .

5. Muscles are connected to bones by _____ .

6. Skeletal and _____ muscle looks _____ under a microscope.

7. Under a microscope, digestive muscles look _____ .

8. Muscles can only _____ bones; they cannot _____ bones.

9. Skeletal muscles act in _____ .

10. When one skeletal muscle _____ , its partner muscle

_____ .

Match each term in Column A with its description in Column B. Write the correct letter in the space provided.

Column A

_____ 1. cardiac muscle

_____ 2. smooth muscle

_____ 3. skeletal muscle

_____ 4. striated

_____ 5. tendons

Column B

a) voluntary muscle

b) found in blood vessels

c) striped

d) heart muscle

e) connect skeletal muscle to bones

What are nutrients?

KEY TERM

nutrient: chemical substance in food needed by the body for growth, energy, and life processes

LESSON 5 | What are nutrients?

You must eat to stay alive. Food supplies you with certain important chemicals called **nutrients** [NOO-tree-unts]. Your body needs nutrients for growth and energy.

There are five groups of nutrients. Most foods supply several nutrients. Most foods, however, are very rich in one or two nutrients.

All the nutrients work together to keep you in good health. Life cannot go on without these nutrients.

CARBOHYDRATES

Carbohydrates [kar-buh-HY-drayts] supply energy. There are two kinds of carbohydrates: sugars and starches.

FATS

Fats also supply energy. But it is usually stored energy. In addition, fats help keep the body warm.

PROTEINS

Proteins [PRO-teenz] are needed to build and heal tissue. Protein is also an important part of **protoplasm** [PROHT-uh-plaz-um], the living material of cells.

VITAMINS

Vitamins [VYT-uh-minz] help control chemical reactions in the body. For example, vitamins control the amount of energy that cells give off. Vitamins also are needed for proper growth.

MINERALS

Minerals are important for healthy tissue. For example, minerals build strong bones and teeth. Muscles, nerves, and blood also need minerals.

Some animals eat only plants. These animals are called <u>herbivores</u> [HUR-buh-vawrz].

Some animals eat only meat (other animals). Animals that eat only meat are called <u>carnivores</u> [KAR-nuh-vawrz].

Some animals eat plants and meat. An animal that eats plants and meat (other animals) is called an <u>omnivore</u> [AHM-nuh-vawr]. Bears, mice, and birds are examples of omnivores. So are humans.

Think of all the things you eat. Do you eat only plants? Do you eat only meat? Probably not. You obtain your nutrients from both plants and other animals.

Ten common foods are listed in the chart. Some of these foods come only from plants. Some come only from animals. Others are mixtures of plant and animal products. You decide where each food comes from.

COMPLETE THE CHART

*Mark a **P** next to the foods that come only from plants. Mark an **A** next to the foods that come only from animals. Mark a **P/A** next to foods that are mixtures of both plants and animals. Then, fill in the third column about each food.*

	FOOD	P, A or P/A	Do you eat this food?
1.	bread		
2.	corn		
3.	lamb chops		
4.	clam chowder		
5.	steak		
6.	pancakes		
7.	beef stew		
8.	milk		
9.	eggs		
10.	tuna salad		

Answer these questions.

1. What do we call an animal that eats only plants? _____

2. What do we call an animal that eats only animals? _____

3. What do we call any animal that eats both plants and animals?

_____

THE IMPORTANCE OF WATER

You have just learned that nutrients are very important. Another substance is also vital to life. It is water. In fact, water is one of the most important substances. You could live for a few months without food. You could live only a few days without water. Why is water so important?

- Our cells are mostly water.

- The life functions cannot take place without water.

How do you get water? Of course you can drink it, but all foods also contain water. Some foods have a great amount of water. Others have only a little water. We can find out if a food contains water by performing a simple test.

TESTING FOR WATER

What You Need (Materials)

test tube and holder
food to be tested (pieces of fruit or
    vegetables or any other food)
Bunsen burner

How To Do The Experiment (Procedure)

1) Place the food into the test tube.

2) Heat gently over Bunsen burner. Be sure to tilt the tube away from you.

Moisture on the inside of the test tube near the top means that the food has water.

What You Learned (Observations)

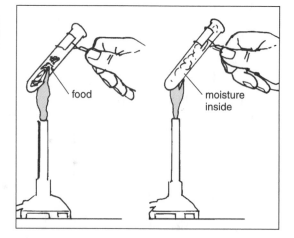

Figure A

Answer these questions about the test for water.

1. Did moisture form on the inside of the tube? _____

2. Do you think the moisture came from the food or the air? _____

Something To Think About (Conclusion)

Does the food that you tested have water? _____

The graph below shows, in percent, how much water there is in some foods. Study the graph for a few minutes. Then answer the questions below.

percent of water in foods

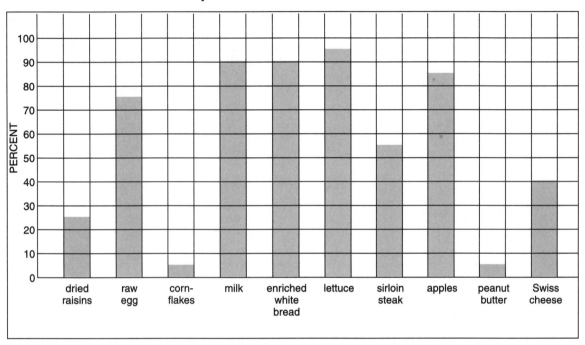

1. What percentage of water does each of these foods have?

   a) dried raisins _____

   b) raw egg _____

   c) cornflakes _____

   d) milk _____

   e) enriched white bread _____

   f) lettuce _____

   g) sirloin steak _____

   h) apples _____

   i) peanut butter _____

   j) Swiss cheese _____

2. Which of these foods has the most water? _____

3. Which of these foods has the least water? _____

Figure B

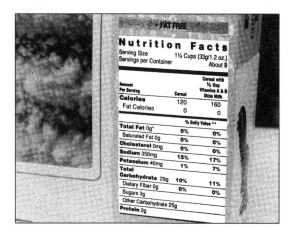

Figure C

Every packaged (canned, boxed, or frozen) food has a label. That's the law! The label lists the ingredients found in the food. The ingredients are listed in order or quantity (amounts). The ingredient found in the greatest amount is listed first. The ingredient with the smallest amount is listed last.

The main contents of a popular cereal are listed below. Study the label. Then answer the questions.

Figure D

1. The ingredient found in the greatest amount is _____ .

2. The ingredient found in the least amount is _____ .

3. This cereal tastes good to many people.

   a) Why do you think many people like the taste? _____

   b) Do you think this is a healthy cereal? _____
      yes, no

   Why? _____

What are carbohydrates, fats, and proteins?

KEY TERMS

carbohydrate: nutrient that supplies energy

fat: energy-storage nutrient

protein: nutrient needed to build and repair cells

amino acid: building block of proteins

LESSON 6 | What are carbohydrates, fats, and proteins?

CARBOHYDRATES

Make a list of the foods you eat in one day. Chances are that about half your diet is made up of **carbohydrates**. That's about normal for most Americans.

What are carbohydrates? Carbohydrates are chemical compounds. They are made up of only one carbon, hydrogen, and oxygen—in certain proportions (balanced amounts).

There are two groups of carbohydrates—starches and sugars. Starches and sugars are "energy" foods. During digestion, starches and double sugars are changed to glucose. Glucose is the simple sugar our bodies "burn" during respiration. This supplies the energy we need to carry out the life processes.

FATS

Like carbohydrates, **fats** are "energy" nutrients. In fact, fats provide more than twice the energy of an equal weight of carbohydrates.

Fats can be either solids or liquids. Solid fats come mostly from animals. Liquid fats are called **oils**. Like carbohydrates, fats are made of carbon, hydrogen, and oxygen.

Fats are very important. They cushion the body and give it shape. Every cell membrane contains fat.

Our bodies contain fat tissue. Important nutrients are stored in this tissue. Fat also helps to insulate the body against the cold.

PROTEINS

Proteins are the building blocks of living matter.

The body uses proteins in several ways. The two most important uses of proteins are:

- to build new cells, and
- to repair damaged cells.

What is the chemical make-up of proteins? Proteins contain atoms of carbon, hydrogen, oxygen, and nitrogen. Some proteins also contain sulfur.

What You Need (Materials)

small pieces of apple (or any other fruit)
Benedict's solution
test tube and holder
Bunsen burner

How To Do The Test (Procedure)

1) Place a few small pieces of the apple in the test tube.

2) Add Benedict's solution (make the test tube about one-third full).

3) Place the test tube over a flame so that the bottom of the tube just touches the flame. Tilt the test tube so that it is pointed away from you.

4) Boil the mixture for about one minute. BE CAREFUL!

If the Benedict's solution turns orange or brick red, then simple sugar is present. If it turns a darker orange, then a lot of simple sugar is present. A light greenish color means very little sugar is present.

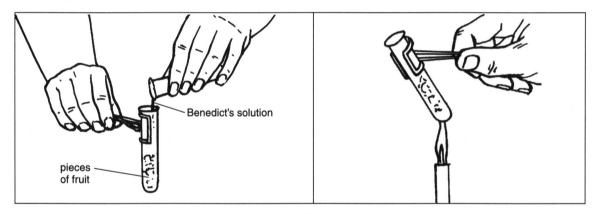

Figure A   **Figure B**

What You Learned

Answer the following questions about the test for sugar.

1. Did the mixture change color? _____

2. What color did the mixture become? _____

3. Does the fruit tested have simple sugar in it? _____

4. What is the name of the special chemical that we used to test for simple sugar?

_____

TESTING FOR STARCH

What You Need (Materials)

Slice of bread or potato
iodine or Lugol's solution
dropper

How To Do The Test (Procedure)

1. Place a drop of iodine or Lugol's solution on the food. The food will turn blue-black if it has starch.

What You Learned (Observations)

Answer the following questions about the test for starch.

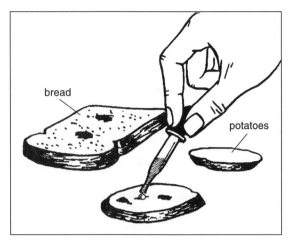

Figure C

1. Did the bread or potato turn blue-black? _____

2. Does the bread or potato have starch? _____

3. What liquid did you use to test for starch? _____

TESTING FOR FATS

What You Need (Materials)

butter (or margarine)
piece of brown wrapping paper

How To Do The Test (Procedure)

1. Rub a small amount of the butter on the paper. Fat makes a spot on the paper. Light can pass through the paper at that spot. The oil makes the paper <u>translucent</u> [trans-LOO-sent].

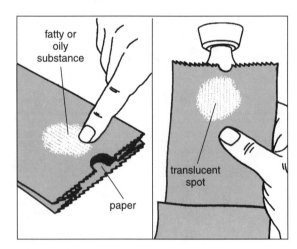

Figure D

What You Learned (Observations)

Answer these questions about the test for oil.

1. Did the butter (or margarine) make a spot on the paper? _____

2. Did the oil make the paper translucent? _____

3. Does the material you tested contain oil? _____

Proteins are made up of smaller compounds called **amino acids**. Amino acids can link up in many different ways. Because of this, there are many kinds of proteins.

Your body uses twenty different amino acids. It can make 12 of them. The other eight must come from food.

- When proteins are digested, the amino acids break away from one another.

- The blood carries the amino acids to the cells. The cells put the amino acids together. They become proteins again.

There are thousands of kinds of proteins. Different cells need different kinds of proteins. Each cell "custom makes" the proteins it needs.

Proteins are giant molecules. They are very complicated. A single protein molecule may have as many as 100,000 amino acids. That is large as far as molecules go. Yet a protein is still very tiny. You cannot see a protein molecule even with the most powerful microscope.

HOW THE BODY MAKES PROTEINS

Look at Figure E and F and read about them. Then answer questions 1–7 on the next page.

Each one of these shapes stands for an amino acid. There are twenty different amino acids.

Amino acids link up to make proteins. Different kinds of link-ups make different kinds of protein.

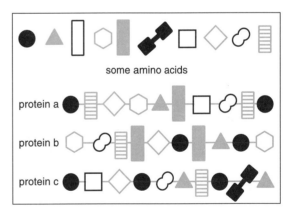

Figure E

1. Food contains proteins. We eat food.

2. Digestion separates the amino acids in the food protein. Blood sends the amino acids to every cell in the body.

3. The cells put these amino acids together again. They become proteins. The cells also put together the amino acids made by the body.

The body needs thousands of different kinds of proteins. Each cell makes the kinds that it needs.

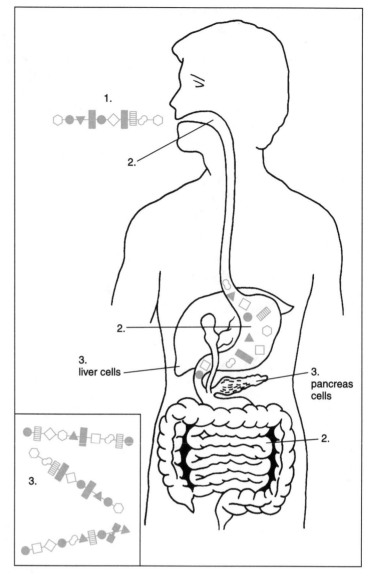

Figure F

ANSWER THESE QUESTIONS

1. How many kinds of amino acids are there? _____

2. What is built when amino acids link up? _____

3. Where do we get proteins? _____

4. What does digestion do to the protein we eat? _____

5. How do amino acids get to cells in every part of the body? _____

6. What do the cells do with the amino acids? _____

FILL IN THE BLANK

Complete each statement using a term or terms from the list below. Write your answers in the spaces provided.

digestion   oxygen   twelve
hydrogen   twenty   starches
amino acids   sugars   carbon
eight   respiration   twice
liquid

1. Carbohydrates are compounds made up of only _____,

   _____, and _____.

2. The two kinds of carbohydrates are _____ and _____.

3. The "burning" of a "fuel" by a cell to obtain energy is called _____.

4. Fats supply more than _____ the amount of energy of carbohydrates.

5. Oils are _____ at room temperature.

6. Proteins are built from linked-up chemicals called _____.

7. The number of amino acids is _____.

8. The number of amino acids a person's body can make is _____.

9. The number of amino acids that we must get from foods is _____.

10. Proteins are broken down into amino acids during _____.

MATCHING

Match each term in Column A with its description in Column B. Write the correct letter in the space provided.

	Column A	Column B
_____	1. glucose	a) liquid fats
_____	2. eight amino acids	b) main jobs of fats
_____	3. cell repair and building new cells	c) simple sugar
_____	4. oils	d) cannot be made by the body
_____	5. cushion, insulate, give shape, and store certain nutrients	e) main jobs of proteins

SCIENCE *EXTRA*

Replacing Damaged Skin

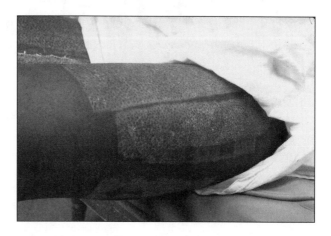

Your skin has probably recovered from many small scrapes, cuts, bruises, and burns. Usually, skin that is damaged can heal itself. But skin cancer and third-degree burns can cause more serious damage. When skin is badly damaged, it may not mend on its own. Doctors have developed a technique called skin grafting that can replace badly damaged skin.

First, the doctor uses very sharp instruments to cut a piece of healthy skin from one part of a person's body. That part of the body will form scars. For this reason, skin used for grafts is usually taken from areas that can be covered by clothing. The healthy skin is moved to an area where the skin has been destroyed.

The doctor then carefully places the healthy skin where it is needed. To hold the grafted skin in place temporarily, doctors use stitches, staples, or bandages. While the skin is healing, it must be kept clean and protected from infection. If the operation is successful, the skin will attach itself to the new site in a matter of weeks. Once attached, it will grow and become a permanent part of the body.

If a large area of skin has been burned, doctors may have trouble finding enough healthy skin for the grafts. In those cases, small pieces of healthy skin are stretched to cover a larger area. The skin is stretched by making slits in the skin and pulling it to make a mesh-like layer. The gaps in the skin heal in time.

Artificial skin is made from chemicals and animal tissue. It can be used to cover a wound temporarily. It can also be used to help the body grow new skin and to form a base for skin grafts.

While a cut or small burn can heal by itself, serious burns need more help. Skin grafting is a way to help the body continue its own healing. The result is new skin that can perform the important jobs that skin has to do.

Lesson **7**

What are vitamins and minerals?

KEY TERMS

vitamin: nutrient found naturally in many foods

mineral: nutrient needed by the body to develop properly

LESSON 7 | What are vitamins and minerals?

There are two additional nutrients that your body needs to carry out its life processes that have not yet been explained. These nutrients are **vitamins** [VYT-uh-minz] and **minerals**. Your body needs small amounts of vitamins and minerals so it can function properly. Just what are vitamins and minerals?

VITAMINS   Vitamins are nutrients that are found naturally in many foods. Most of the vitamins the body needs are found in foods. However, two vitamins, vitamin D and vitamin K are made in your body.

Vitamins are important for:

• Maintaining proper growth.

• Helping to keep bones and teeth strong.

• Helping to keep muscles and nerves healthy.

• Helping to change carbohydrates and fats to energy.

MINERALS   Minerals are nutrients needed by the body to develop properly. You need large amounts of some minerals. You need small amounts of other minerals. Each mineral has a different job. For example:

• Iron is needed to form red blood cells.

• Calcium and **phosphorus** [FAHS-fur-us] are needed to keep bones and teeth strong.

• Sodium is needed for healthy muscles and nerves.

• Chlorine is needed to produce an enzyme needed for digestion.

• Iodine controls body growth.

If your body does not get enough of the vitamins and minerals it needs, a deficiency disease may develop. Vitamins, minerals, and deficiency diseases will be discussed further in this lesson.

Vitamins are needed for good health. Most foods contain several vitamins. But some foods are very rich in one or more vitamins. The pictures below show six groups of foods. The foods in each group are extra rich in one particular vitamin.

Figure A *Important sources of vitamin A*

Figure B *Important sources of vitamin C*

Figure C *Important sources of vitamin B*

Figure D *Important sources of vitamin D*

Figure E *Important sources of vitamin K*

Figure F *Important sources of vitamin E*

Your body needs small amounts of vitamins every day. If you do not get enough of a certain vitamin, you can develop a deficiency disease. Night blindness is caused by a deficiency of vitamin A. Look at the table to see other vitamin deficiency diseases.

VITAMIN	USE IN BODY	SOURCES	DEFICIENCY DISEASE
A	healthy skin, eyes, ability to see well at night, healthy bones and teeth	orange and dark green vegetables, eggs, fruit, liver, milk	night blindness
B_1 thiamin	healthy nerves, skin, and eyes; helps body get energy from carbohydrates	liver, pork, whole grain foods	beriberi
B_2 riboflavin	healthy nerves, skin, and eyes; helps body get energy from carbohydrates, fats, and proteins	eggs, green vegetables, milk	skin disorders
B_3 niacin	works with B vitamins to get energy from nutrients in cells	beans, chicken, eggs, tuna	pellagra
C	healthy bones, teeth, and blood vessels	citrus fruits, dark green vegetables	scurvy
D	healthy bones, teeth; helps body use calcium	eggs, milk, made by skin in sunlight	rickets
E	healthy blood and muscles	leafy vegetables, vegetable oil	none known
K	normal blood clotting	green vegetables, tomatoes	poor blood clotting

Study the chart below. It lists several minerals, their importance, what foods they are found in, and signs of a deficiency disease.

Figure G   *Food rich in potassium*   **Figure H**   *Food rich in magnesium*

MINERAL	USES	SOURCES	SIGNS OF DEFICIENCY
calcium	builds bones and teeth	milk and milk products, canned fish, green leafy vegetables	soft bones poor teeth
phosphorus	builds bones and teeth	red meat, fish, eggs, milk products, poultry	none known
iron	builds red blood cells	red meat, whole grains, liver, egg yolk, nuts, green leafy vegetables	anemia (paleness, weakness, tiredness) brittle fingernails
sodium	helps keep muscles and nerves healthy	table salt, found naturally in many foods	none known
iodine	used to make a chemical that controls releases of energy by cells	seafood, iodized salt	goiter
potassium	helps keep muscles and nerves healthy	bananas, oranges, meat, bran	loss of water from cells, heart problems, high-blood pressure
magnesium	strong bones and muscles, nerve action	nuts, whole grains, green leafy vegetables	none known
zinc	formation of proteins that speed up reactions	milk, eggs, seafood, milk, whole grains	none known

43

Answer each of the questions. Look back at the chart for the answers. Search carefully and be patient and you will get them all correct. But remember, this is make-believe. In real life, never diagnose or treat your own health problems. SEE YOUR DOCTOR!

1. James is always tired. He lacks energy and looks pale.

   a) Which mineral deficiency disease might James have?

   _____

   b) Which mineral might James be lacking? _____

   c) What food or foods might help this problem?

   _____

Figure I

2. Ann, too, is sluggish and tired. She also has a poor appetite. Sometimes Ann's muscles cramp.

   a) What vitamin deficiency disease might Ann have?

   _____

   b) Which missing vitamin might be the cause? _____

   c) What food or foods might help Ann's problem?

   _____

Figure J

3. It is a sunny day. Tom goes into a movie house. He stumbles around looking for a seat. Things look very dark to him for several minutes.

   a) Which vitamin deficiency disease might Tom have? _____

   b) Which nutrient might be missing from his diet? _____

   c) What foods should Tom eat to solve this problem? _____

MATCHING

Match each vitamin to the deficiency disease it is most closely associated with. Write the correct letter in the space provided.

Deficiency Disease   **Vitamin**

_____ 1. night blindness   **a.** vitamin D

_____ 2. beriberi   **b.** vitamin C

_____ 3. scurvy   **c.** vitamin A

_____ 4. pellagra   **d.** vitamin B_3

_____ 5. rickets   **e.** vitamin B_1

_____ 6. poor blood clotting   **f.** vitamin K

COMPLETE THE CHART

Complete the chart. You may write more than one mineral name in the last column.

	Use	**Mineral(s)**
1.	Builds strong bones and teeth	
2.	Helps keep muscles and nerves healthy	
3.	Formation of red blood cells	
4.	Makes chemical that controls release of energy by cells	
5.	Formation of proteins that speed up reactions	

Complete each statement using a term or terms from the list below. Write your answers in the spaces provided. Some words may be used more than once.

night blindness	iodine	vitamins
rickets	K	deficiency disease
iron	D	muscles
nerves	minerals	

1. Nutrients that are found naturally in food are _____ .

2. A _____ is caused by a diet that is missing a certain nutrient.

3. An example of a deficiency disease that causes soft bones is _____ .

4. Anemia can result from too little _____ .

5. Sodium is needed for healthy _____ and _____ .

6. _____ are nutrients, such as calcium and iron.

7. Two vitamins that can be made by the body are _____ and

   _____ .

8. If your blood does not clot properly, you may be deficient in vitamin

   _____ .

9. Seafood and iodized salt are good sources of the mineral _____ .

10. A deficiency of vitamin A could result in _____ .

A man visits his doctor. He has symptoms of weak bones and weak teeth. The doctor suspects that the patient's diet does not supply enough calcium and phosphorus. However, tests show a normal level for each of these nutrients. What vitamins might this man be

deficient in? _____

What is a balanced diet?

LESSON 8 | What is a balanced diet?

Many people overeat. Yet they are poorly fed. Eating a lot does not always mean that we are eating properly.

A diet must be balanced. A balanced diet supplies proper amounts of all the nutrients. You can plan a balanced diet by including foods from the four basic food groups. All foods have been classified into four groups. The four groups are:

DAIRY GROUP

This food group includes milk and milk products. Cheese, butter, and yogurt are examples of dairy products.

BREAD AND CEREAL GROUP

This food group includes whole grain or enriched grain products. Bread, cereal, rice, crackers, and pastas are examples of foods in this group.

MEAT GROUP

This food group includes meat, fowl, fish, and eggs. Nuts, peas, and beans that are high in protein are also included in this group.

VEGETABLE AND FRUIT GROUP

This food group includes green and yellow vegetables, citrus fruits, tomatoes, bananas, and raisins.

You should eat foods from each food group each day. The total number of servings you eat from each group is called your daily intake.

Every meal should include at least one protein. Proteins include eggs, meat, fish, fowl, and milk products. At least one citrus fruit should be eaten every day.

Take advantage of what you have learned. Learn to choose foods wisely from these groups. Then you will be sure of having a balanced diet. Eating properly is an important key to good health.

Fill in the chart below with four specific examples of foods that belong in each food group. Some foods are shown in Figure A to help you.

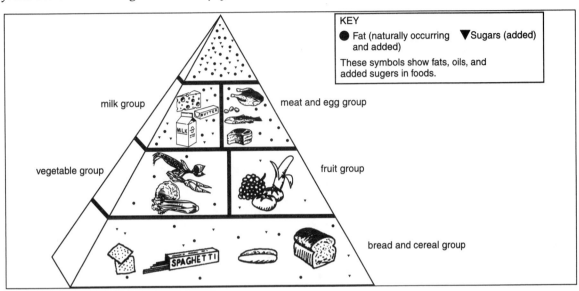

KEY
● Fat (naturally occurring ▼Sugars (added)
and added)
These symbols show fats, oils, and
added sugers in foods.

milk group
meat and egg group
vegetable group
fruit group
bread and cereal group

Figure A

	DAIRY (MILK) GROUP	MEAT GROUP	FRUIT AND VEGETABLE GROUP	BREAD AND CEREAL GROUP
1.				
2.				
3.				
4.				

FILL IN THE BLANK

Complete each statement using a term or terms from the list below. Write your answers in the spaces provided.

meat group   vegetable and fruit group protein
nutrients   dairy group   balanced
bread and cereal group

1. The useful things we get from food are called _____ .

2. A proper diet is called _____ .

3. Foods can be placed into four groups. They are the _____ ,

   the _____ , the _____ , and the

   _____ .

4. Nuts and beans are part of the meat group because they are high in _____ .

TRUE OR FALSE

In the space provided, write "true" if the sentence is true. Write "false" if the sentence is false.

_____	**1.** A person who eats a lot always has a balanced diet.
_____	**2.** A balanced diet supplies the right amounts of all the nutrients.
_____	**3.** A person who eats only meat has a balanced diet.
_____	**4.** Eggs can take the place of meat in the diet.
_____	**5.** Bread and cereals supply some protein.
_____	**6.** Some vegetables are high in vitamin C.
_____	**7.** White bread supplies the same nutrients as citrus fruits.
_____	**8.** Ice cream contains milk.
_____	**9.** There are six basic food groups.
_____	**10.** How we eat can affect our health.

REACHING OUT

Use what you have learned to plan three well balanced meals: breakfast, lunch, and supper. List the foods that you would include in each meal in the spaces below. Remember to include at least one protein in each meal.

Breakfast	Lunch	Supper

How is food digested?

KEY TERMS

digestion: process by which foods are changed into forms the body can use

esophagus: tube that connects the mouth to the stomach

peristalsis: wavelike movement that moves food through the digestive tract

LESSON 9 | How is food digested?

People, like all living things, need food. Food gives us the nutrients our bodies need. It also gives us energy. Energy is needed to carry out the life processes.

Our bodies cannot use the nutrients or energy in food unless the food is changed. The changing of food into a form the body can use is called **digestion** [dy-JES-chun].

What does digestion do? Digestion breaks down large pieces of food into smaller pieces. Digestion also changes the chemicals of food. It changes large, complex food molecules into smaller, simpler ones.

Where does digestion take place? Digestion takes place in the <u>digestive tract</u>. The digestive tract is a long, curving tube in your body. If stretched out, the digestive tract would be more than 9 meters (30 feet) long.

What are the parts of the digestive tract? The parts of the digestive tract are: the mouth, the **esophagus** [i-SAF-uh-gus], the stomach, the small intestine, and the large intestine.

There are many glands and organs along the digestive tracts, such as the liver and the pancreas. These organs are not part of the digestive tract, but they help in digestion. The digestive tract and the other digestive organs make up your digestive system.

Food enters the body through the mouth. Waste materials (undigested food) leave the body through the **anus** [AY-nus]. The anus is at the end of the large intestine.

Digestion is a step-by-step process. It does not take place quickly. It takes food from one to two days to pass through the entire digestive tract.

Read the descriptions below to find out what happens as food moves through the digestive tract.

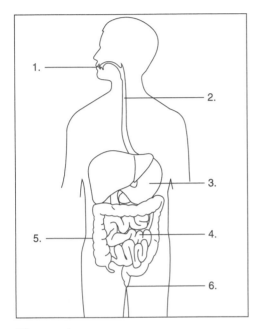

Figure A

1. Food enters the body through the mouth. Digestion begins here.

   • The teeth break food into smaller pieces.
   • Saliva moistens the food.
   • Saliva also begins the chemical breakdown of starch.

2. When you swallow, food passes into the esophagus.

   • Food passes through the esophagus and into the stomach.

3. What happens to food in the stomach?

   • The stomach churns food and breaks it into even smaller pieces.
   • The chemical digestion of protein begins.
   • Partially digested food then moves to the small intestine.

4. Most digestion takes place in the small intestine.

   • The small intestine is also where all digestion is completed.
   • Undigested food (waste) then passes into the large intestine.

5. Undigested food (waste) is not used by the body.

   • The large intestine stores and eliminates undigested food as solid waste.

6. Solid waste is passed out of the body through the anus.

   • NOTE: The anus is <u>not</u> a digestive organ.

Figure B shows the organs of the digestive system. Study the drawing, then see if you can answer the questions below.

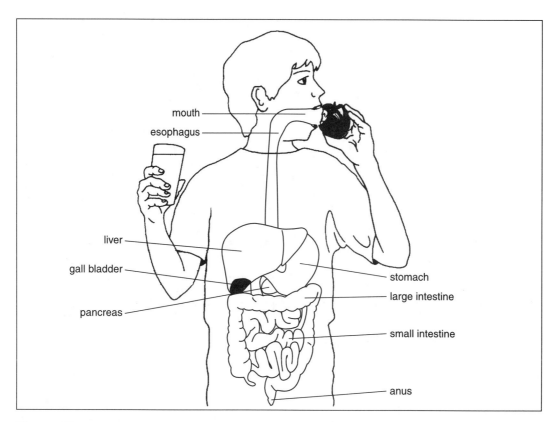

Figure B

1. Name the parts of the digestive tract that food passes through in order. (Do not include the anus.) _____ , _____ , _____ , _____ , and _____ .

2. The digestive tract has two openings to the outside of the body.

   a) Food enters the body through the _____ .

   b) Waste materials leave the body through the _____ .

3. **a)** Where does chemical digestion start? _____

   b) Where does most chemical digestion take place? _____

4. What are two organs that are part of your digestive system, but **not** part of the digestive tract? _____ and _____

FILL IN THE BLANK

Complete each statement using a term or terms from the list below. Write your answers in the spaces provided. Some words may be used more than once.

esophagus   stomach   smaller
completed   simpler   small intestine
digestive tract   teeth   saliva
large intestine   mouth   digestion
liver   pancreas

1. The changing of food to a form the body can use is called _____.

2. Digestion breaks down large pieces of food into _____ pieces.

   Digestion also makes food molecules _____.

3. Digestion takes place in a body tube called the _____.

4. The parts of the digestive tract (in order) through which food passes are the

   _____, the _____, the _____, the

   _____, and the _____.

5. In the mouth, food is broken into smaller pieces by the _____.

6. Food in the mouth is moistened by _____.

7. Starch digestion starts in the _____.

8. Protein digestion starts in the _____.

9. Most digestion takes place in the _____. This is also where all

   digestion is _____.

10. Organs such as the _____ and _____ help in digestion,

    but are outside the digestive tract.

HOW FOOD MOVES ALONG THE DIGESTIVE TRACT

Food in the digestive tract does not move by itself. Food is squeezed along the digestive tract by wavelike movements of muscles. These muscles work by themselves. We do not have to think about moving food. It is an involuntary action. The wavelike movement is called **peristalsis** [per-uh-STAWL-sis]. Peristalsis starts in the esophagus right after you swallow. It continues along the entire digestive tract. Peristalsis works in only one direction—except when we are ill. For sample, reverse peristalsis in the stomach or esophagus causes us to "throw up." Throwing up, or vomiting, is one way the body gets rid of things that can harm us.

Get a rubber tube. Wet the inside.

Put in a marble that just fits in the tube.

Pinch it forward.

This will give you an idea of how peristalsis moves food through the digestive tract.

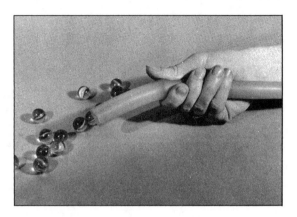

Figure C

IDENTIFY THE PARTS

Identify the parts of the digestive tract by writing the letter from Figure D next to the name of the part.

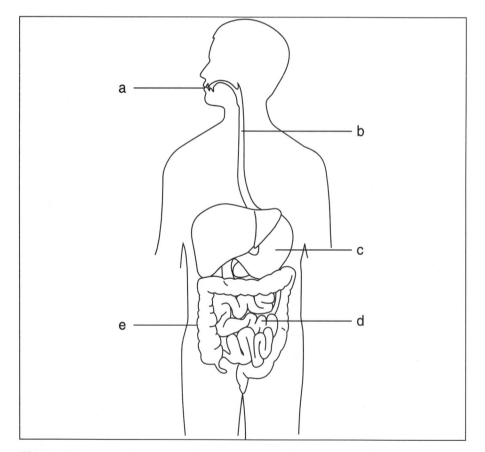

Figure D

1. small intestine _____   3. mouth _____   5. esophagus _____

2. stomach _____   4. large intestine _____

56

How do enzymes help digestion?

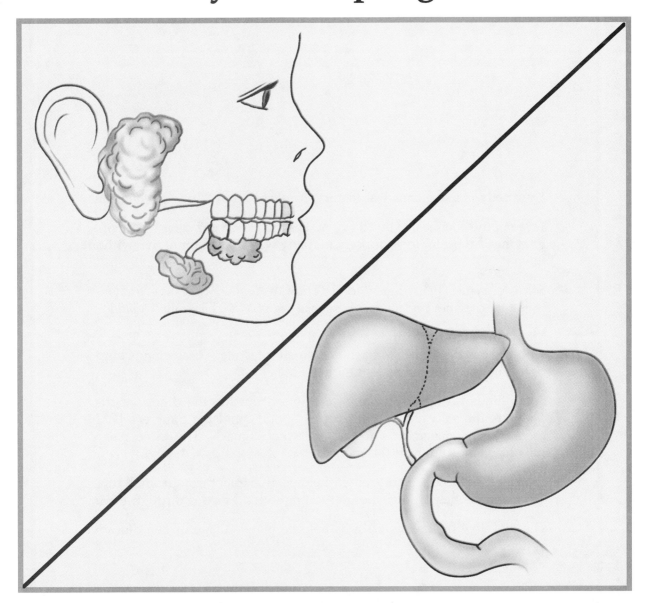

KEY TERMS

enzyme: protein that controls chemical activities

bile: green liquid that breaks down fats and oils

LESSON 10 | How do enzymes help digestion?

Your body is like a chemical factory. It makes many kinds of chemicals.

Some of the chemicals your body makes are called **enzymes** [EN-zymz]. Enzymes are useful to your body in many ways. You cannot live without them.

Some enzymes help digest food. They are called digestive enzymes. Digestive enzymes are made by special groups of cells called glands.

Many tiny digestive glands are found inside the digestive tract. They are within the walls of the stomach and small intestine. These glands empty right into the stomach and small intestine.

Some kinds of digestive glands are found outside the digestive tract. They are the **salivary** [SAL-uh-ver-ee] glands, and the **pancreas** [PAN-kree-us]. The pancreas is found near the stomach. The three pair of salivary glands are near the mouth.

Enzymes from these glands enter the digestive tract through small tubes. These glands help in digestion although no food passes through them.

The digestive tract along with its helping glands make up the digestive system.

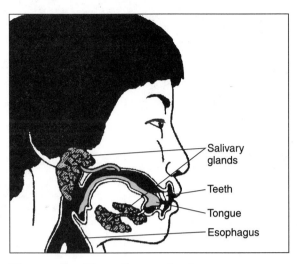

Figure A   *The salivary glands*

THE SALIVARY GLANDS

The salivary glands produce saliva. Saliva is mostly water. It also contains an enzyme called **ptyalin** [TY-uh-lin].

Water in saliva moistens food. This makes the food easier to swallow.

Ptyalin starts changing starch to simple sugars.

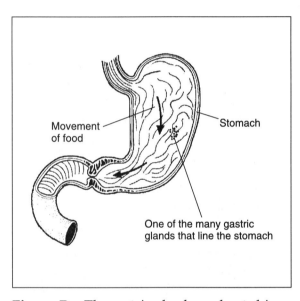

Figure B   *The gastric glands are located in the walls of the stomach.*

THE STOMACH

The digestive glands of the stomach are called gastric glands. These glands secrete (give off) a liquid called gastric juice.

Gastric juice contains the enzymes **pepsin** [PEP-sin] and rennin. It also contains **hydrochloric** [hy-druh-KLAWR-ik] acid and mucus.

- Pepsin starts protein digestion.

- Rennin "curdles" milk. It changes liquid milk protein to a "cheeselike" substance. This keeps the protein from passing through the digestive tract too quickly. It gives protein-digesting enzymes time to digest the protein.

- Hydrochloric acid. Pepsin can digest protein properly only in an acid environment. Hydrochloric acid, in the stomach, provides this environment. The mucus in gastric juice helps protect the stomach lining from the acid.

The pancreas and the small intestine produce enzymes. These enzymes complete the digestion of all nutrients.

Look at Figure C carefully. Then complete the sentences below.

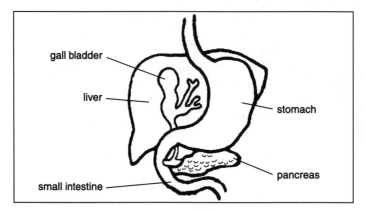

Figure C

1. Enzymes from the pancreas empty into the _____ .

2. The pancreas lies just below the _____ .

THE LIVER

The liver is one of the largest organs in the human body. It also helps in digestion. The liver provides a liquid called **bile**. Bile is not an enzyme. But it is very important in fat digestion. Bile breaks fat into tiny pieces. It "prepares" the fat for fat-digesting enzymes.

Bile does not move directly from the liver to the small intestine. It is stored in the gall bladder. When you eat fat, the gall bladder squeezes. Some bile is forced out of the gall bladder. It goes into the small intestine. Bile mixes with the food in the small intestine.

The chart below tells the story of how digestion in the small intestine changes the chemicals of food:

STARTING PRODUCT		END PRODUCT
starches and double sugars	→ change to	simple sugars
proteins	→ change to	amino acids
fats	→ change to	simpler fatty substances

MATCHING

Match each term in Column A with its description in Column B. Write the correct letter in the space provided.

Column A

_____ **1.** liver

_____ **2.** enzymes

_____ **3.** glands

_____ **4.** digestive system

_____ **5.** mouth

Column B

a) chemicals made by the body

b) starting point of digestion

c) make body chemicals

d) one of the largest organs

e) digestive tract and other digestive organs

LABEL THE DIAGRAM

Find each part of the digestive system and write its letter in the blank.

1. stomach _____

2. large intestine _____

3. mouth _____

4. pancreas _____

5. small intestine _____

6. esophagus _____

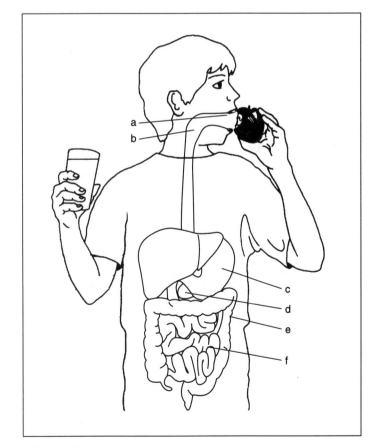

Figure D

FILL IN THE BLANK

Complete each statement using a term or terms from the list below. Write your answers in the spaces provided.

simple sugars	ptyalin	liver
small intestine	fats	gall bladder
outside	smaller molecules	large molecules
salivary glands	glands	starch
digestive enzymes	pancreas	stomach

1. Any chemical digestion changes _____ to

    _____ .

2. Enzymes are produced by groups of cells and tissues called _____ .

3. Chemicals that help digest food are called _____ .

4. Digestive glands are found both within and _____ the digestive tract.

5. Digestive glands within the digestive tract are found in the walls of the

    _____ and the _____ .

6. Two digestive glands that are found outside the digestive tract are the

    _____ and the _____ .

7. The enzyme found in saliva is _____ .

8. Ptyalin starts to change _____ to _____ .

9. Bile is produced by the _____ . It is stored in the _____ .

10. Bile breaks up _____ .

END PRODUCTS, PLEASE

Fill in the correct answers.

1. The end product of fat digestion is _____ .

2. The end product of starch digestion is _____ .

3. The end product of double sugar digestion is _____ .

4. The end product of protein digestion is _____ .

PROVING THAT PEPSIN NEEDS HYDROCHLORIC ACID TO WORK

What You Need (Materials)

three small pieces of egg white, hard-boiled
pepsin liquid
2% hydrochloric acid
water
three test tubes and rack

How To Do The Experiment (Procedure)

1. Place one piece of egg white in each test tube. Label the test tubes A, B, and C.

2. To test tube A, add the pepsin liquid until it is one-quarter full. Then add water until it is half full.

3. To test tube B, add the same amount of pepsin. Then carefully add hydrochloric acid until the test tube is half full.

4. To test tube C, add water until it is half full. This test tube is our control. The control allows us to compare what happens to the egg white with and without the enzymes.

Put the test tubes in the rack. Let them stand overnight. (See Figure F.)

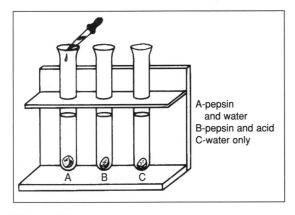

A-pepsin
  and water
B-pepsin and acid
C-water only

Figure E

Figure F *24 hours later*

What You Learned (Observations)

1. The egg white in test tube C _____ change.
    <small>did, did not</small>

2. The egg white in test tube B has changed _____ .
    <small>only slightly, a great deal</small>

3. The egg white in test tube A has changed _____ .
    <small>only slightly, a great deal</small>

4. a) The egg white has almost completely dissolved in test tube _____ .
A, B, C

b) Most of the egg white in this test tube has changed to a _____ .
solid, liquid

5. In which test tube has no digestion taken place? _____
A, B, C

6. In which test tube has only slight digestion taken place? _____
A, B, C

7. In which test tube has the most digestion taken place? _____
A, B, C

Something To Think About (Conclusions)

1. Water _____ digest protein.
does, does not

2. Pepsin alone _____ digest protein, but very _____ .
does, does not   slowly, fast

3. a) Pepsin digests protein quickly when it is mixed with an _____ .
(one word)

b) Name the acid in gastric juice. _____

4. Chemical digestion changes _____ molecules into _____ molecules.
large, small   large, small

REACHING OUT

Sometimes a gall bladder becomes diseased. It has to be taken out. People can live without their gall bladders, but their eating habits must change. How would they have to change their diets?

_____

_____

_____

_____

What is absorption?

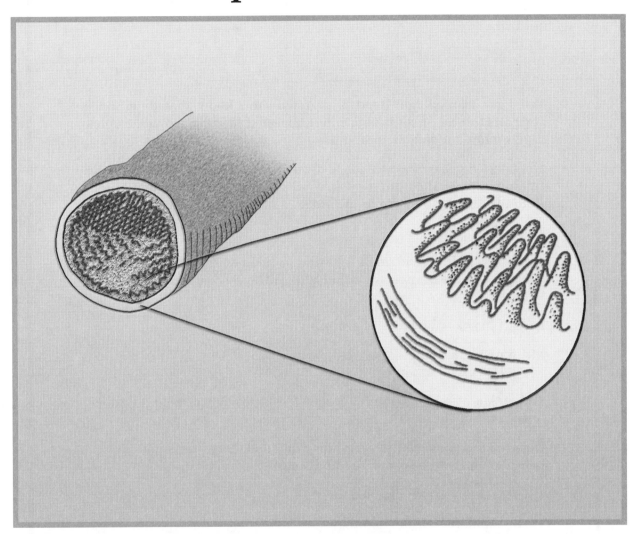

KEY TERMS

absorption: movement of food from the digestive system to the blood

villi: fingerlike projection on the lining of the small intestine

LESSON 11 | What is absorption?

You have learned how food is digested. You know that digestion is completed in the small intestine. But, digested food has no value unless it gets into your cells.

How does digested food leave the digestive system? It is absorbed in the small intestine. **Absorption** [ab-SAWRP-shun] is the movement of food from the digestive system to the blood.

This is how it happens:

The inner wall of the small intestine is lined with thousands of tiny "bumps." These bumps are called **villi** [VIL-y]. (One bump is called a villus.)

Each villus has two kinds of tubes:

1. a network of capillaries, and
2. a lacteal.

As you know, capillaries carry <u>blood</u>. The lacteals carry a liquid called <u>lymph</u>.

Digested food surrounds each villus. The food leaves the small intestine through the capillaries and lacteals.

• The lacteals absorb digested <u>fats</u>.

• The capillaries absorb <u>all</u> other nutrients.

Lymph and blood flow through the body in separate tubes. But the two liquids do not stay separated. Lymph empties into the bloodstream near the heart. Then, the blood carries all the nutrients.

As you know, blood goes to every part of the body. The cells absorb the nutrients from the blood.

Study Figures A and B.

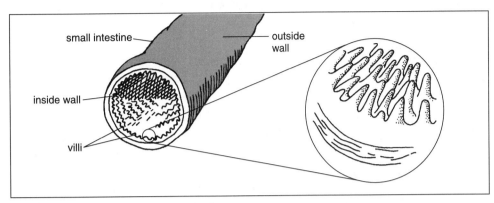

Figure A   *The inner wall of the small intestine has thousands of tiny villi.*

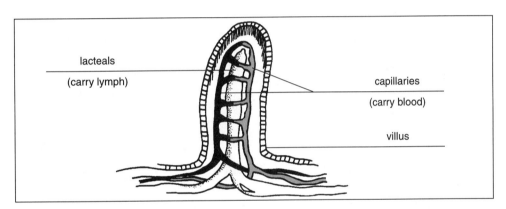

Figure B   *A single villus*

1. Digested fats are absorbed by the _____ .
    capillaries, lacteals

2. Digested starches, sugars, and proteins are absorbed by the _____ .
    capillaries, lacteals

3. Digested food is carried to all parts of the body by the _____ .
    blood, small intestine

MATCHING

Match each term in Column A with its description in Column B. Write the correct letter in the space provided.

Column A	Column B
_____ **1.** digestion	**a)** lined with villi
_____ **2.** lymph	**b)** changes food to a form the body can use
_____ **3.** small intestine	**c)** absorb all digested food except fats
_____ **4.** capillaries	**d)** lacteal liquid

67

FILL IN THE BLANK

Complete each statement using a term or terms from the list below. Write your answers in the spaces provided. Some words may be used more than once.

mouth	cells	fats
lacteal	small intestine	nutrients
all digested nutrients	digestion	capillaries
absorption	villi	

1. Our bodies are made up of trillions of _____ .

2. All the useful things we get from food are called _____ .

3. The breakdown of food into simpler and smaller molecules is called

   _____ .

4. Digestion starts in the _____ and is completed in the

   _____ .

5. The movement of food from the digestive system to the blood is called

   _____ .

6. Absorption of food takes place through tiny "bumps" called _____ .

7. Villi line the inner wall of the _____ .

8. Every villus has _____ and a _____ .

9. Lacteals absorb digested _____ .

10. Capillaries in villi absorb _____ except fats.

REACHING OUT

1. Which covers more surface area, a flat surface or a bumpy surface?

   _____

2. How does the shape of the villi speed absorption?

   _____

Lesson **12**

What is the circulatory system?

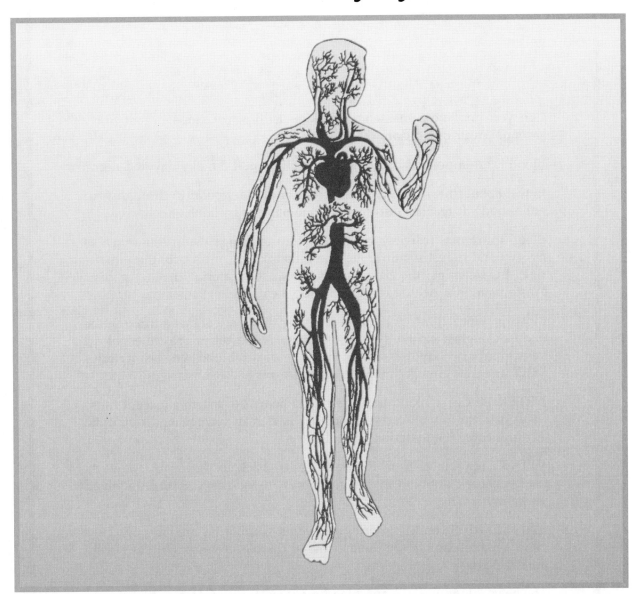

KEY TERMS

arteries: blood vessels that carry blood away from the heart

veins: blood vessels that carry blood to the heart

capillaries: tiny blood vessels that connect arteries to veins

LESSON 12 | What is the circulatory system?

Can you imagine a messenger making trillions of stops in just thirty seconds? Your blood does!

Blood is an important body messenger. It is on the move day and night.

In just about thirty seconds, your blood moves (circulates) through your entire body. It reaches out to every one of your trillions of cells.

Blood transports (carries) to the cells all the things they need—such as oxygen and digested nutrients. The cells take in, or absorb, these materials. In exchange, the blood picks up waste materials from the cells. Waste materials include carbon dioxide, heat, and extra water.

Blood is pumped throughout the body by the heart. It flows through the body through a closed system of tubes. These tubes are called blood vessels. Your body has three main types of blood vessels: **arteries** [ART-ur-ees], **veins** [VANES], and **capillaries** [KAP-uh-ler-ees].

ARTERIES Carry blood away from the heart. All arteries (except those that go to the lungs) carry blood that is rich in oxygen and nutrients. Arteries carry the materials the cells need.

VEINS carry blood from the body (cells) back to the heart. All veins (except those coming from the lungs) carry blood that contains dissolved waste materials.

CAPILLARIES connect arteries and veins. Capillaries are very tiny. You need a microscope to see them. Most of the blood vessels in your body are capillaries.

The heart, blood vessels, and blood make up the **circulatory** [SUR-kyuh-luh-towr-ee] system. Circulation, or transport, is a vital function. Life cannot go on without it.

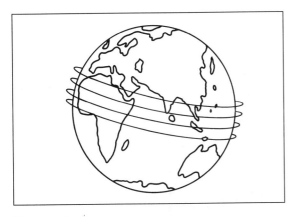

Figure A

Blood vessels are found in almost every part of the body.

If laid end to end, your blood vessels would stretch out to about 161,000 kilometers (100,000 miles)!

That's about 4 times the distance around the equator!

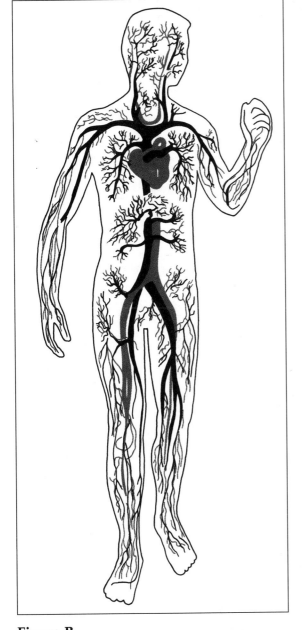

Look at Figure B. The grey tubes show the arteries. The black tubes show the veins. Many thousands of tiny capillaries connect the arteries and the veins.

Write the correct term in each blank to answer the questions or complete the sentence.

1. What pumps blood through your

   body? _____

2. Blood vessels that carry blood away from the heart are called

   _____ .

3. Vessels that carry blood back to the

   heart are called _____ .

4. Blood moves from arteries to veins through tiny blood vessels called

   _____ .

5. The heart, blood vessels, and blood

   make up the _____ .

Figure B

Look at Figure C. Then answer these questions:

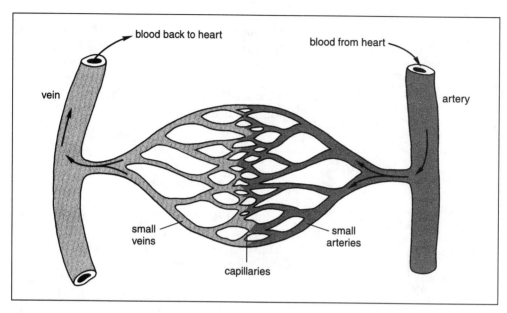

Figure C

1. Arteries branching away from the heart become _____ .
   _{smaller and smaller, larger and larger}

2. Veins leading back to the heart become _____ .
   _{smaller and smaller, larger and larger}

3. Most of our blood vessels are _____ .
   _{arteries, veins, capillaries}

MATCHING

Match each term in Column A with its description in Column B. Write the correct letter in the space provided.

Column A	Column B
_____ **1.** circulation	**a)** carry blood away from the heart
_____ **2.** heart	**b)** pumps blood
_____ **3.** arteries	**c)** connect arteries and veins
_____ **4.** veins	**d)** transport of materials in living things
_____ **5.** capillaries	**e)** carry blood back to the heart

FILL IN THE BLANK

Complete each statement using a term or terms from the list below. Write your answers in the spaces provided. Some words may be used more than once.

heart
capillaries
nutrients
veins

circulation
arteries
waste materials

oxygen
blood vessel
blood

1. The transport of materials in living things is called _____ .

2. In humans, circulation is carried out by the liquid called _____ .

3. Blood is pumped by the _____ .

4. Blood brings to cells things like _____ and _____ .

5. Blood picks up _____ from the cells.

6. Any tube that carries blood is called a _____ .

7. The three kinds of blood vessels are _____ , _____ , and

_____ .

8. Blood is carried away from the heart by _____ .

9. Blood is carried back to the heart by _____ .

10. Arteries and veins are connected by tiny blood vessels called _____ .

WORD SCRAMBLE

Below are several scrambled words you have used in this Lesson. Unscramble the words and write your answers in the spaces provided.

1. UCATLOCNIRI _____

2. NIVE _____

3. ILYCARLAP _____

4. THERA _____

5. RATYER _____

TRUE OR FALSE

In the space provided, write "true" if the sentence is true. Write "false" if the sentence is false.

_____ 1. Circulation is the transport of materials in living things.

_____ 2. Life stops when circulation stops.

_____ 3. Blood is pumped by the brain.

_____ 4. Blood circulates through the body only a few times a day.

_____ 5. Arteries carry blood away from the heart.

_____ 6. Arteries transport carbon dioxide to the cells.

_____ 7. Veins carry blood away from the heart.

_____ 8. Capillaries pick up waste materials from the cells.

_____ 9. Capillaries connect arteries and veins.

_____ 10. Capillaries are the largest blood vessels.

REACHING OUT

Circulation is always carried out by a liquid. In humans and many other animals, that liquid is blood.

What liquid do you think carries out circulation in plants? _____

_____

What is blood made of?

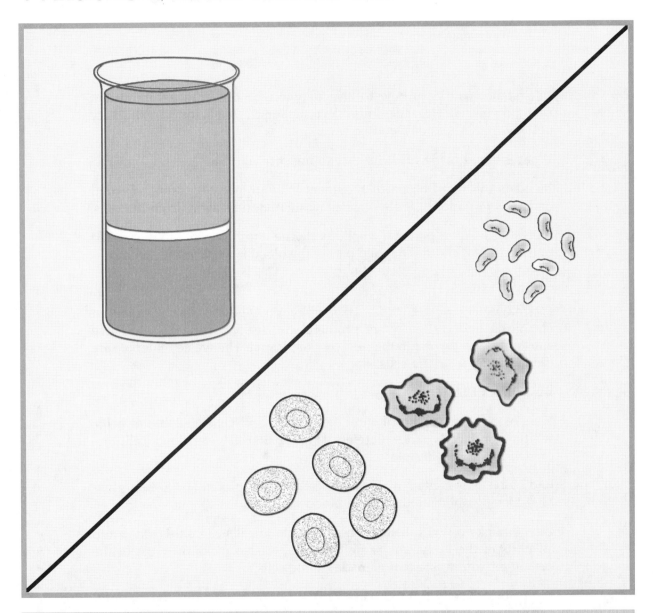

KEY TERMS

plasma: liquid part of blood

red blood cells: cells that give blood its red color and carry oxygen

white blood cells: cells that protect the body from disease

platelets: tiny, colorless pieces of cells help blood to clot

LESSON 13 | What is blood made of?

In first aid, you learn how to help people who are hurt. An important first-aid rule is: "Treat Serious Bleeding First." A person can die very quickly from a loss of blood.

What is blood made of? Why is it so important to life?

Blood is the main tissue of transport in your body. It carries needed materials to the cells. It also carries waste materials away from the cells.

Blood has a liquid part and a solid part. The <u>liquid</u> part of blood is called **plasma** [PLAZ-muh]. The <u>solid</u> part of the blood is made up of blood cells.

PLASMA

Plasma is 90% water. It is straw colored. Digested food, important chemicals, and certain waste products are dissolved in plasma. These substances are carried to the cells by the plasma. The waste materials are carried away from the cells.

BLOOD CELLS

Your blood is made up of three kinds of blood cells: **red blood cells, white blood cells**, and **platelets** [PLAYT-lits]. These blood cells are carried in the flowing plasma.

<u>Red</u> <u>blood</u> <u>cells</u> contain a substance called hemoglobin [HEE-moh-gloh-bin]. Hemoglobin is red. It gives blood its color.

Oxygen links up with hemoglobin. Red blood cells carry this oxygen to all parts of the body. The same hemoglobin also picks up most of the carbon dioxide waste that is made by the cells.

<u>White</u> <u>blood</u> <u>cells</u> fight disease and infection. They destroy harmful germs in the body.

<u>Platelets</u> are tiny, colorless pieces of cells. They help stop bleeding. Platelets give off a chemical that helps blood clot.

BLOOD COMPOSITION

Figure A shows the composition of blood. Study Figure A, then answer the questions.

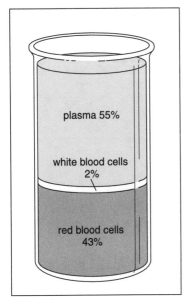

Figure A *Make-up of blood*

1. What percent of blood is liquid?

    _____

2. What is the name of the liquid part

    of blood? _____

3. **a)** The liquid part of blood is made

    up mostly of _____ .
    (If you need help, look back to
    the reading.)

    b) What percent? _____

4. All of the blood cells together make

    up _____ percent of
    blood.

5. Red blood cells make up _____ percent of blood; white blood cells

    make up _____ percent.

BLOOD CELLS—THEIR SIZES AND NUMBERS

Figure B will give you an idea of the sizes and numbers of red and white blood cells found in your body. Study Figure B. Then answer the questions.

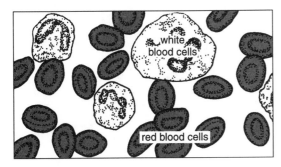

Figure B

1. Which blood cells are the largest? _____

2. Which type of blood cell is most numerous? _____

3. Which cells are shaped like "pinched" disks? _____

Study Figures C and D. Then answer the questions about each picture.

4. What kind of blood cell is shown?

   _____

5. Describe briefly what is happening

   in Figure C. _____

   _____

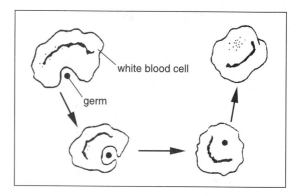

Figure C *A white blood cell at work.*

Look at Figure D. Then answer the questions.

6. When you cut yourself, which part of the blood helps you stop bleeding?

   _____

7. White blood cells also come to the area of a cut. Why? _____

8. Take a guess! What happens to the number of white blood cells when germs are in

   the blood? _____

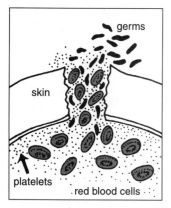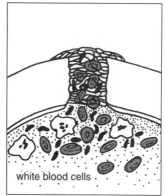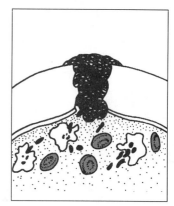

Figure D *Blood clots when your skin is cut.*

Answer the following questions about red blood cells.

1. Oxygen is _____ .
   needed by the cells, a cell waste

2. Which blood cells pick up and carry oxygen? _____
   red blood cells, white blood cells, platelets

3. What substance in red blood cells joins with oxygen? _____

4. Where does the blood pick up this oxygen? _____

   a) in the heart

   b) in the arteries and veins

   c) in the lungs

5. What gives blood its color? _____

TO OR AWAY?

Blood has been called the "River of Life." Blood carries to the cells materials the cells need. In turn, blood carries away waste materials made by the cells.

Nine substances carried by the blood are listed in the chart below. Indicate whether each substance is carried to the cells or away from the cells. Place a check (✔) in the proper boxes.

	SUBSTANCE CARRIED BY THE BLOOD	CARRIED TO THE CELLS	CARRIED AWAY FROM THE CELLS
1.	digested food		
2.	oxygen		
3.	carbon dioxide		
4.	enzymes		
5.	hormones (used by the cells to regulate chemical reactions)		
6.	heat		
7.	harmful chemicals		
8.	extra (waste) water		
9.	vitamins and minerals		

1. Blood makes up about 9% of a person's body weight. For example, if you weigh 100 pounds, 9 pounds is blood. (Figure out how many pounds of blood your body has.)

2. An adult has about 12 pints of blood.

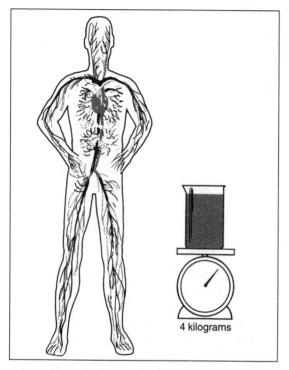

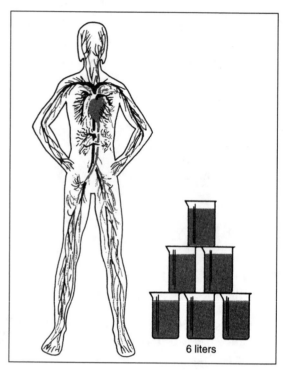

Figure E   *Blood makes up about 9% of a person's weight. If you weigh 45 kilograms (100 pounds), 4 kilograms (9 pounds) is blood.*

Figure F   *A grownup has about 5.7 liters (12 pints) of blood.*

3. There are about 600 red blood cells for every white blood cell in your blood. Just one drop of blood contains about 5 million red blood cells. There are about 25 trillion red blood cells in the body of the average adult.

4. Red blood cells and white blood cells are produced in the marrow of bones—especially in the ribs, breastbone, and backbone.

5. It is estimated that from 1 to 2 million red blood cells die every <u>second</u>. New cells are made to take their place.

6. Blood cells are carried by the flowing plasma. White blood cells, however, can also move by <u>themselves</u>.

7. White cells also can pass through tiny holes in the blood vessels. They move into surrounding tissues. White blood cells are like good soldiers. They seek out and destroy enemies (harmful germs).

How does your heart work?

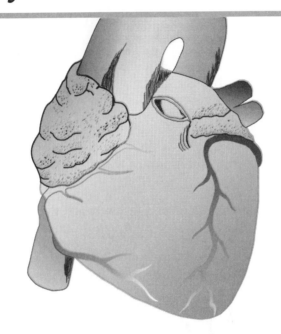

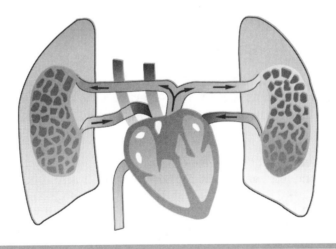

KEY TERMS

atria: upper chambers of the heart

ventricles: lower chambers of the heart

valve: thin flap of tissue that acts like a one-way door

septum: thick tissue wall that separates the left and right sides of the heart

LESSON 14 | How does your heart work?

Place your hand on your chest. The beating you feel is from your heart. It is keeping you alive. Your heart is mostly muscle tissue. It has only one job. Day and night, twenty-four hours a day, your heart pumps blood to every part of your body.

The human heart is divided into four separate spaces called chambers. Two chambers are in the upper part of the heart; two chambers are in the lower part.

ATRIA The upper chambers of the heart are the right and left **atria** [AY-tree-uh]. The singular of atria is atrium. Atria receive blood.

- The right atrium receives blood from all parts of the body. Blood in the right atrium is high in carbon dioxide and low in oxygen.

- The left atrium receives blood from the lungs. Blood in the left atrium is high in oxygen and low in carbon dioxide.

Both atria fill with blood at the same time.

VENTRICLES The lower chambers of the heart are the **ventricles** [VEN-tri-kuls]. Ventricles pump blood out of the heart.

- The right ventricle pumps blood to the lungs. This blood is high in carbon dioxide and low in oxygen. When the blood is in the lungs, it gives up its carbon dioxide. At the same time, the blood picks up fresh oxygen.

- The left ventricle pumps blood to all parts of the body except the lungs. Blood in the left ventricle is high in oxygen. It is low in carbon dioxide.

Both ventricles pump blood out of the heart at the same time. Every time your heart beats, blood is being "squeezed" out of the ventricles.

Blood moves in only one direction. The heart and veins have **valves** that keep the blood from flowing backward. A valve is a thin flap of tissue.

A muscular wall divides the right side of the heart from the left side. This wall is called the **septum**. Blood cannot flow from one side of the heart to the other.

The human heart is like two separate pumping systems. One system serves the lungs. The other, serves the entire body.

Let us trace the path of blood into and out of the heart. **NOTE**: The heart diagrams are shown as if you were looking at the front of a person. The right side of the heart appears on the left side of the drawings. The left side of the heart appears on the right.

REMEMBER: In a working heart, both upper chambers fill with blood at the same time. Both lower chambers squeeze (pump blood out) at the same time.

We will study the right side of the heart first. Then, we will study the left side of the heart. In this way, you will better understand how the circulatory system works.

Veins Carry Blood from all Parts of the Body to the Heart

Answer the questions below. Search the reading and study the diagrams carefully to find the answers.

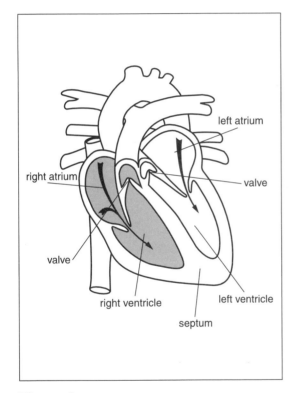

Figure A

1. Which chamber receives blood from all the body veins?

   _____

2. **a)** The blood passes from this chamber into the

   _____ .

   b) While this is happening, the valve between the right atrium and right ventricle is

   _____ .
   open, closed

3. The blood in the right ventricle is

   high in _____ , and low
   oxygen, carbon dioxide

   in _____ .
   oxygen, carbon dioxide

4. The body _____ use this blood.
   can, cannot

5. Where must this blood go to get a fresh supply of oxygen? _____

The Right Ventricle Contracts. It Squeezes Its Blood Out of the Heart to the Lungs.

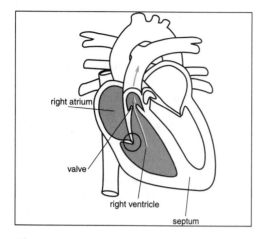

Figure B

6. **a)** When the right ventricle contracts, the valve between the upper and lower chambers is

   _____ .
   <u>open, closed</u>

   b) What does this prevent? _____

   _____

7. Blood pumped out of the right ventricle goes _____ .
   <u>to the body, to the lungs</u>

8. In the lungs, the blood gives up its _____ , and picks up
   <u>oxygen, carbon dioxide</u>

   _____ .
   <u>oxygen, carbon dioxide</u>

9. The blood _____ be used by the cells.
   <u>can, cannot</u>

10. Where must the blood go before it can be sent to the entire body?

    _____

Veins Carry Fresh Blood Back to the Heart from the Lungs.

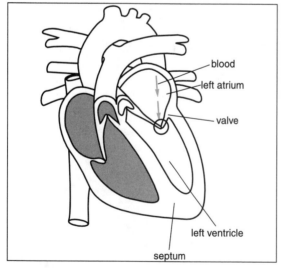

Figure C

11. Which chamber receives fresh blood from the lungs? _____
    <u>left atrium, left ventricle</u>

12. **a)** The blood then passes into the

    _____ .
    <u>left atrium, left ventricle</u>

    b) While this is happening, the valve between the left chambers

    is _____ .
    <u>open, closed</u>

The Left Ventricle Contracts. This Forces Blood Out of the Heart to all Parts of the Body Except the Lungs.

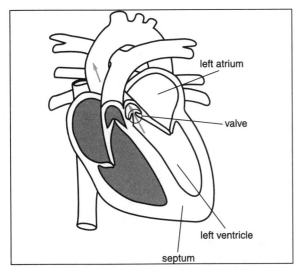

13. a) When the left ventricle contracts, the valve between the left

chambers is _____ .

open, closed

b) Why? _____

_____

Figure D

14. Where does blood leaving the left ventricle go? _____

_____

15. THEN, what do you think happens to the blood? _____

WHAT IS A PULSE?

Every time your ventricles contract, blood is forced out of your heart and into your arteries. This force pushes blood through your arteries in spurts. With each spurt, a beat can be felt. This beat is called a pulse.

Each pulse beat tells you that your ventricles are contracting.

PULSE BEAT = HEARTBEAT

How fast does your heart beat? It depends on several things—like age, activity, and how calm or excited you are.

The heart of a rested adult beats about 70 times a minute. A young person's heart beats slightly faster.

Activity, fear, worry, and excitement are all things that make the heart beat faster.

You can feel a pulse on an artery that is near the skin.

Each pulsebeat tells us that blood is being pumped out of the heart.

There are several pulse spots. Most pulses are taken on the wrist. Figure E shows how to do it. See if you can feel your pulse. (Do not use your thumb.)

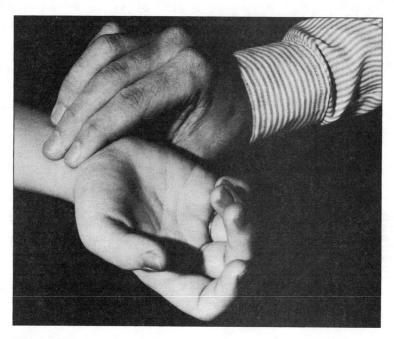

Figure E

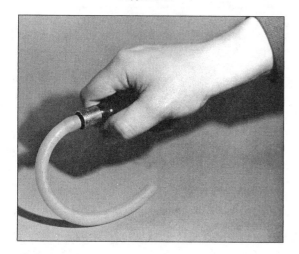

Figure F

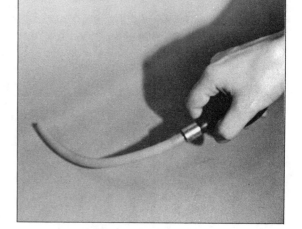

Figure G

You can get an idea how a pulse works.

Squeeze a rubber bulb filled with water. The tube connected to it will "jump." This "jump" is like a single pulsebeat.

TAKING PULSE RATES

What You Need (Materials): yourself

How To Do the Experiment (Procedure)

1. Take your pulse at a resting state.

2. Run in place for 30 seconds. Take your pulse again.

What You Learned (Observations)

1. Fill in your results in the chart below.

	Name	Test #1 Rested Pulse Rate	Test #2 Pulse Rate After Exercise
1.			
2.			
3.			
4.			
5.			
	Totals		
	AVERAGES (Add all the pulses in each test group. Then divide by 5.)		

Something To Think About (Conclusions)

1. Everyone's pulse rate _____ the same.
   <u>is, is not</u>

2. Exercise makes the pulse beat _____ .
   <u>slower, faster</u>

WORD SCRAMBLE

Below are several scrambled words you have used in this Lesson. Unscramble the words and write your answers in the spaces provided.

1. SSEVLE _____

2. TRIUMA _____

3. PTUSEM _____

4. CLETRINEV _____

TRUE OR FALSE

In the space provided, write "true" if the sentence is true. Write "false" if the sentence is false.

_____ 1. The heart is a muscle.

_____ 2. The heart has many jobs.

_____ 3. A human heart has four chambers.

_____ 4. Heart chambers are called arteries and veins.

_____ 5. Blood moves from the atria to the ventricles.

_____ 6. Ventricles receive blood from veins.

_____ 7. Arteries carry blood away from the heart.

_____ 8. The right and left ventricles pump at the same time.

_____ 9. Your heart stops beating when you are asleep.

_____ 10. Your heart beats millions of times a year.

REACHING OUT

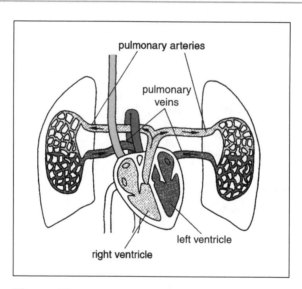

Figure H

1. Arteries carry "fresh" blood. There is one exception. Which artery is the exception?

   _____

2. Veins carry "stale" blood. There is one exception. Which veins are the exception?

   _____

What is breathing and respiration?

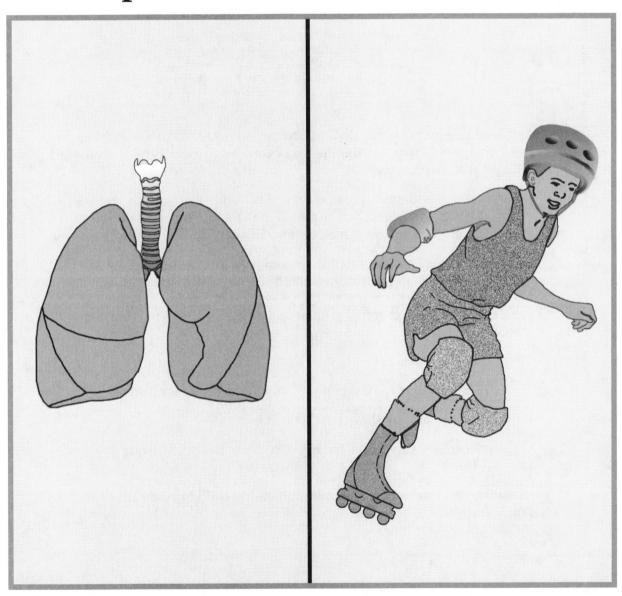

KEY TERM

respiration: process of carrying oxygen to cells, getting rid of carbon dioxide, and releasing energy

LESSON 15 | What is breathing and respiration?

You need energy to live. So do birds, trees, and bacteria. All living things need energy to carry out the life processes. And, there can be no life without the life processes.

How do plants and animals get energy? The same way your car gets its energy, by burning a fuel. Cars use gasoline as a fuel. Energy is released when oxygen from the air combines with the gasoline in the engine.

Animals get energy by linking the oxygen they breathe in with the food that they eat. This important life process is called **respiration** [res-puh-RAY-shun]. Respiration is the energy-producing process in living things. It is the release of energy by combining oxygen with digested food (glucose).

Here is what happens:

Digested Food + Oxygen → Energy and Waste Products

Respiration can also be shown in this way:

Glucose + Oxygen → Energy + Water + Carbon Dioxide
(fuel)   (waste)   (waste)

In humans and many other animals, breathing is done by means of the lungs. Breathing in is taking air into the lungs. Breathing out is forcing the air out of your lungs.

Breathing and respiration are related—but they are not the same. Breathing is necessary for respiration to take place. Breathing is the mechanical process of taking oxygen into the body and sending carbon dioxide out of the body.

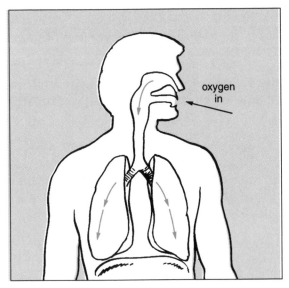

Figure A

Breathing in (inhaling) sends oxygen into the lungs.

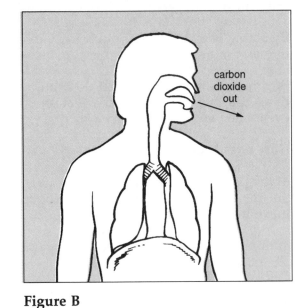

Figure B

Breathing out (exhaling) sends carbon dioxide waste out of the lungs.

Respiration takes place in every cell of the body. Respiration uses the oxygen that inhaling brings into the body.

Try to answer these questions about respiration.

1. What brings the oxygen to all parts

   of the body? _____

2. What does respiration make that

   living things need? _____

3. What waste materials does

   respiration give off? _____

   _____

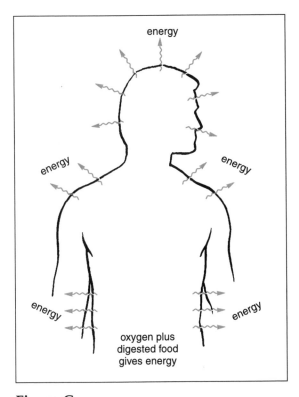

Figure C

Breathing and respiration are related. But they are <u>not</u> the same.

Respiration is a chemical process. It happens in every cell. In respiration, digested food links up with oxygen. This link-up produces the energy the cells need.

Breathing is a mechanical [muh-KAN-ih-cul] action. Breathing is the movement of gases into and out of the lungs.

Breathing is involuntary. You do it automatically without thinking. You breathe all the time. You breathe when you are awake. You breathe when you are asleep. You breathe even when you are unconscious!

How does breathing take place?

Many people believe that air in the lungs makes their chest move in and out when they breathe. This is <u>not</u> true. In fact, the opposite is true. It is your chest size that makes air move in and out of your lungs.

Your chest size changes when you breathe. It changes because of the actions of:

- your rib muscles, and

- your diaphragm [DY-uh-fram] muscle.

INHALING AND EXHALING

Figures D and E show what happens when you breathe. Study the diagrams carefully. Then answer the questions.

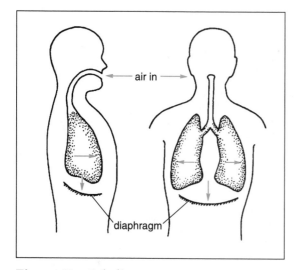

Figure D   *Inhaling*

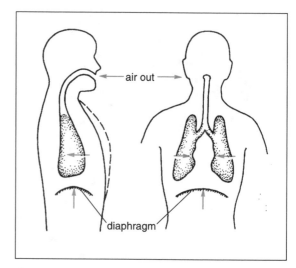

Figure E   *Exhaling*

1. When you inhale (see Figure D),

   a) the ribs move _____ .
      <small>inward, outward</small>

   b) the diaphragm moves _____ .
      <small>upward, downward</small>

   c) there is now _____ space in the chest area.
      <small>more, less</small>

   d) air rushes _____ to fill this space.
      <small>in, out</small>

2. When you exhale (Figure E),

   a) the ribs move _____ .
      <small>inward, outward</small>

   b) the diaphragm moves _____ .
      <small>upward, downward</small>

   c) there is now _____ space in the chest area.
      <small>more, less</small>

   d) because of this pressure, air moves _____ the lungs.
      <small>in, out of</small>

MORE ABOUT BREATHING

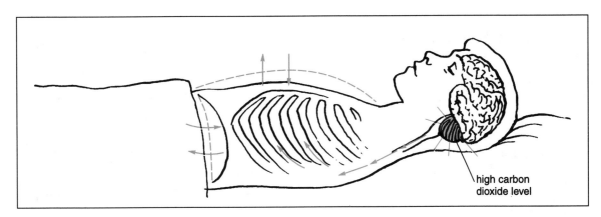

high carbon
dioxide level

Figure F

Why do you breathe?

Breathing is automatic. When the level of carbon dioxide in your blood increases, a message goes to your brain. Then, your brain sends a message to your diaphragm and rib muscles to move. You take a breath without thinking!

INHALING OR EXHALING?

Each of the following goes either with inhaling or exhaling. Place a check (✔) in the box where you think it belongs.

		INHALING	EXHALING
1.	air moves out of the lungs		
2.	air moves into the lungs		
3.	ribs move out		
4.	ribs move in		
5.	chest space becomes smaller		
6.	chest space becomes larger		
7.	diaphragm moves down		
8.	diaphragm moves up		

WORD SCRAMBLE

Below are several scrambled words you have used in this Lesson. Unscramble the words and write your answers in the spaces provided.

1. ALEHEX

2. MUSELC

3. HINALE

4. GMPRHAAID

5. SBRI

What is the respiratory system?

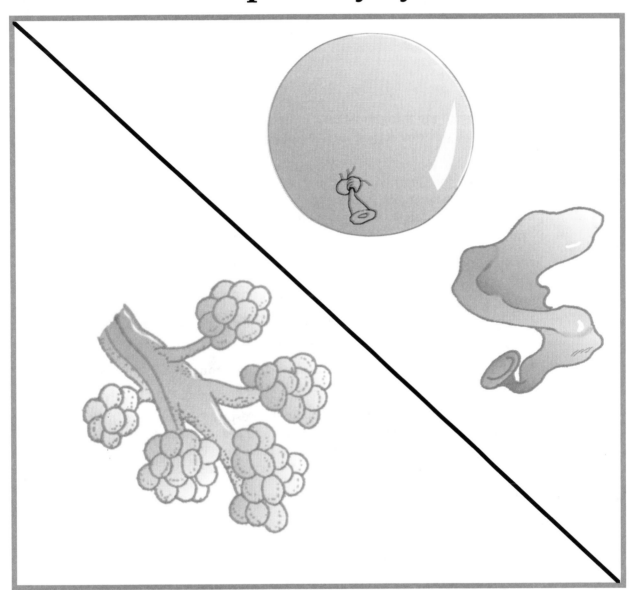

KEY TERMS

trachea: windpipe

bronchi: tubes leading to the lungs

alveoli: microscopic air sacs in the lungs

LESSON 16 | What is the respiratory system?

Almost every living thing must take in oxygen in order to live. Breathing is the process of bringing air into the organism. Breathing also gets rid of used air.

As you have just learned, breathing is done by means of lungs. The lungs, along with several other organs make up the **respiratory** [RES-pur-uh-towr-ee] system. The job of the respiratory system is to take oxygen into the lungs and to get rid of carbon dioxide and water.

Let us trace the path that air takes when you inhale and exhale.

1. Air enters the body through the mouth or nose.

2. The air moves into your throat and then passes through the windpipe, or **trachea** [TRAY-kee-uh].

3. The trachea branches into two tubes called **bronchi** [BRAHN-kee]. Each bronchus extends into one of the lungs.

4. The lungs are the main organs of the respiratory system. In the lungs, the bronchi branch into smaller and smaller tubes. At the end of the smallest tubes are tiny air sacs. Each lung contains millions of air sacs. Each air sac is surrounded by capillaries.

While the air is in the air sacs, two important things happen:

• The blood picks up oxygen from the air sacs.

• At the same time, the air sacs pick up carbon dioxide waste from the blood.

When you exhale, you breathe out the carbon dioxide. Some waste water and heat also are exhaled.

The path that air follows when we breathe is called the respiratory tract. It is shown in Figure A. Study it. Then answer these questions or complete the sentences.

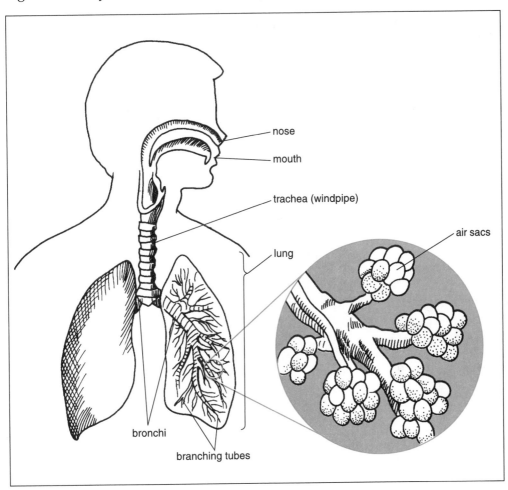

Figure A *Notice the enlarged segment of the lung. Each small branch ends at an air sac.*

1. The respiratory tract starts with the _____ and the _____ .

2. The respiratory tract ends with millions of tiny _____ .

3. How many lungs does a person have? _____

4. The parts of the respiratory tract are listed below. But they are not in order. Rewrite them in the order in which air goes through the body.

   bronchi   mouth and nose   air sacs   trachea   branching tubes

   _____ , _____ , _____ , _____ ,

   _____

5. Each bronchus extends into a _____ .

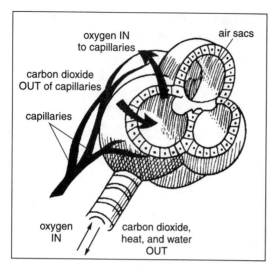

The lungs have millions of air sacs. Air sacs are also called **alveoli** [al-VEE-uh-ly]. Alveoli are very tiny. You need a microscope to see them.

Figure B

1. Air that enters the air sacs is rich in _____ .
   <small>oxygen, carbon dioxide</small>

2. Air that leaves the air sacs is rich in the gas _____ .
   <small>oxygen, carbon dioxide</small>

3. Air sacs are surrounded by _____ .

4. The capillaries around the air sacs take in _____ and give off
   <small>oxygen, carbon dioxide</small>

   _____ .
   <small>oxygen, carbon dioxide</small>

5. List the three waste materials the lungs excrete.

   _____ , _____ , _____

MATCHING

Match each term in Column A with its description in Column B. Write the correct letter in the space provided.

	Column A	**Column B**
_____	1. exhaling	a) where gases are exchanged
_____	2. inhaling	b) windpipe
_____	3. air sacs	c) breathing in
_____	4. trachea	d) surround the air sacs
_____	5. capillaries	e) breathing out

FILL IN THE BLANK

Complete each statement using a term or terms from the list below. Write your answers in the spaces provided. Some words may be used more than once.

alveoli
inhaling
mouth

windpipe
bronchi
exhale

nose
capillaries
smaller and smaller

1. Breathing in is called _____.

2. We inhale through the _____ or _____.

3. The trachea is the scientific name for the _____.

4. The trachea divides into two tubes called _____.

5. In the lungs, the tubes branch into _____ tubes.

6. The lungs have millions of tiny air sacs called _____.

7. Air sacs have many _____.

8. We get rid of carbon dioxide waste when we _____.

LABEL THE DIAGRAM

Identify the parts of the respiratory system. Write the correct letter on the lines provided.

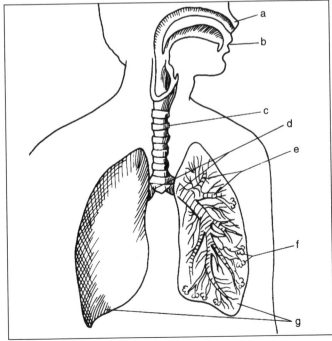

Figure C

1. bronchi ____

2. nose ____

3. branching tubes ____

4. mouth ____

5. air sacs ____

6. trachea ____

7. lung ____

HOW CAN WE SHOW THAT WE EXHALE CARBON DIOXIDE?

What You Need (Materials)

limewater   drinking straw
plastic cup

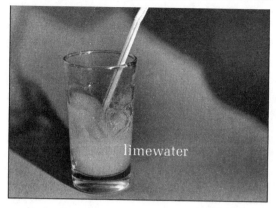

Figure D

What You Need to Know

1. Limewater is a clear liquid.

2. Limewater turns milky when it is mixed with carbon dioxide.

How To Do The Experiment (Procedure)

1. Pour some limewater into the plastic cup.

2. Using the plastic straw, exhale normally into the limewater. **CAUTION: Do not inhale. Inhaling might cause you to suck the poisonous solution into your mouth.**

What You Learned (Observations)

1. The limewater _____ .
   <u>stayed clear, turned milky</u>

Something To Think About (Conclusions)

1. The gases that mixed in with the limewater came from the _____ .
   <u>air, lungs</u>

2. This shows that the lungs release _____ .
   <u>carbon dioxide, oxygen</u>

REACHING OUT

How can you easily show that you also exhale some water and some heat? _____

_____

_____

How does the body get rid of wastes?

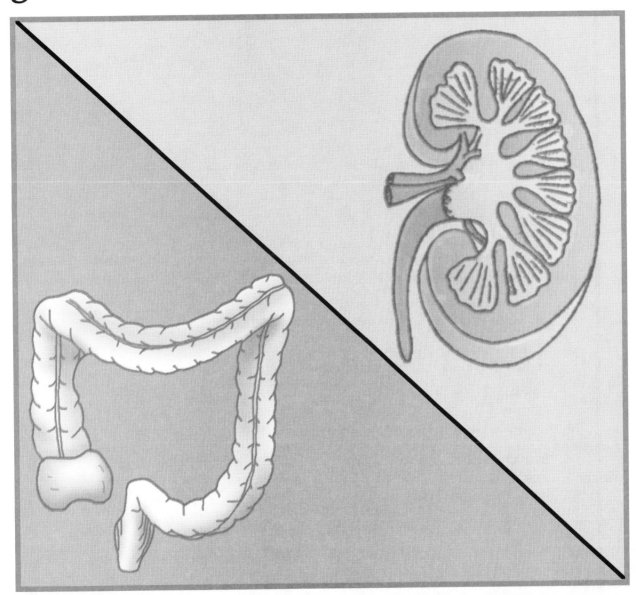

KEY TERMS

elimination: the removal from the body of wastes from digestion

excretion: the removal from the body of wastes made by the cells

LESSON 17 | How does the body get rid of wastes?

Can you imagine a city without sewers, chimneys, or garbage removal? Waste materials would pile up and up. Before long, everyone would have to move. Nobody could live there.

Your body must get rid of wastes too. You cannot live without getting rid of waste products.

The body makes several kinds of waste materials. There are two main groups: undigested solid wastes and wastes made by the cells.

You have learned that undigested solid wastes leave the body through the large intestine. This process is called **elimination**.

The cells make many different wastes. These wastes include water, heat, carbon dioxide, salts, and urea. Urea is a nitrogen compound.

The removal of waste products made by the cells is called **excretion** [ik-SKREE-shun]. In excretion, blood picks up the wastes from the cells. The wastes are sent to special organs that pass them out of the body.

In many animals, carbon dioxide leaves the body through the lungs. The liquid waste, urine, is made in the kidneys. Urine is made up of water, heat, and harmful chemicals. Heat and water, as well as salt, are excreted by the body through the skin as **perspiration** [pur-spuh-RAY-shun].

When glucose from food is combined with oxygen in your cells, heat and other kinds of energy are formed. This energy is used to carry out the life processes. As a result of this process, wastes are formed. This process is shown by the following equation:

$$\textbf{glucose + oxygen} \rightarrow \textbf{carbon dioxide + water + extra heat}$$
$$\textbf{(waste)} \qquad \textbf{(waste)} \quad \textbf{(waste)}$$

If these wastes are not removed from your body, they could prove harmful to you. The organs of the excretory system remove these wastes. For example, the lungs get rid of carbon dioxide and water. The kidneys get rid of liquid wastes. The skin gets rid of liquid wastes and helps you get rid of extra heat.

SOLID WASTES

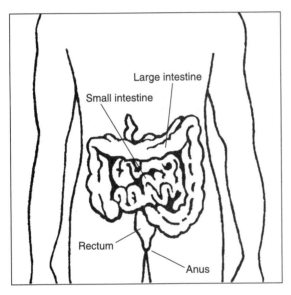

Some parts of food that you eat cannot be digested. This undigested food forms wastes. These wastes move through the small intestine into the large intestine. Water is removed from the wastes in the large intestine. This forms a solid. The solid wastes move from the large intestine into the rectum. From the rectum, the solid wastes are eliminated from the anus.

Figure A

WHAT DOES THE PICTURE SHOW?

Waste materials leave the body through many paths. Look at the picture. Certain parts of the body are labeled. Each part gets rid of certain wastes.

The waste products are:

carbon dioxide   harmful chemicals
solid wastes   water
salt   heat

1. Write the correct waste product next to the part. Remember, some waste materials leave the body through more than one part.

   Skin _____

   Large intestine _____

   Lungs _____

   Kidneys _____

   _____

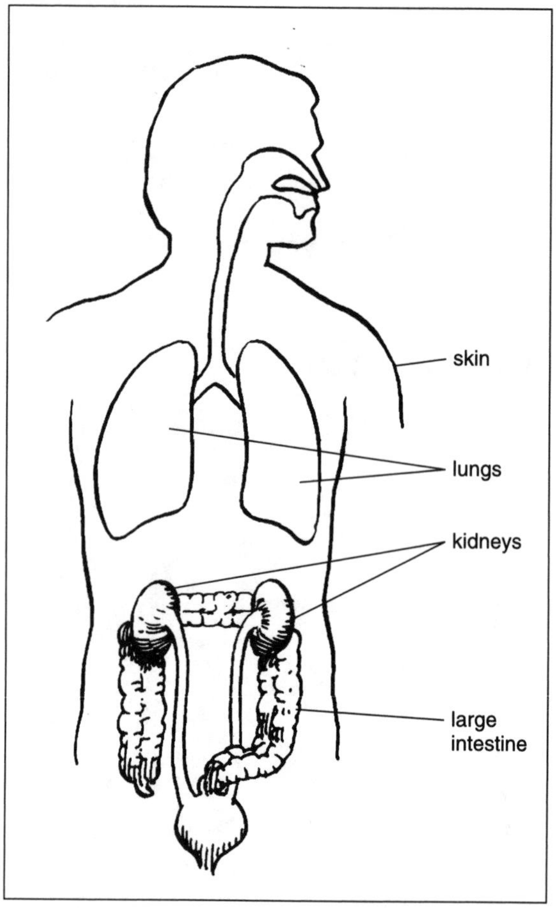

Figure B

REACHING OUT

Why do you think the skin is sometimes referred to as an "air conditioner?" _____

_____

_____

What is the excretory system?

KEY TERM

excretory system: body system responsible for removing cellular wastes from the body

LESSON 18 | What is the excretory system?

Removing waste products from the body is the job of the **excretory system.** The main organs of the excretory system are the lungs, kidneys, and skin.

LUNGS

You have learned that the lungs excrete carbon dioxide waste. The lungs also excrete small amounts of heat and water.

SKIN

The skin excretes most of the body's waste heat. In addition, the skin removes some water, salts, and a very small amount of urea. These wastes are excreted by the skin as sweat or **perspiration** [pur-spuh-RAY-shun]. Perspiration, or sweating, also helps your body "cool off." When sweat evaporates from your skin, it cools the body. Evaporation is the changing of a liquid to a gas. As sweat evaporates, it removes heat from your body.

KIDNEYS

The kidneys excrete a liquid waste called urine. Urine is a mixture. It is made mostly of water and urea. But it also contains some salts. Some heat is also removed from the body by the kidneys.

Figure A shows the organs of the excretory system.

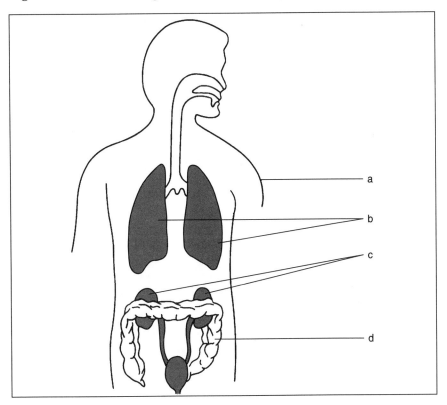

Figure A

1. Can you identify them? Place the correct letter next to the name of each organ on the lines provided.

    kidneys _____   large intestine _____

    lungs _____   skin _____

2. Which of these is an organ of elimination? _____

3. Which are organs of excretion? _____ _____

    _____

4. List five waste products the body must excrete. _____

    _____ _____ _____ _____

5. a) Which one of these waste materials is excreted only by the lungs?

    _____

    b) Which other waste materials do the lungs help excrete? _____

6. The kidneys excrete a liquid mixture called urine.

   a) Name the two main waste products of urine. _____ and

      _____

   b) What other waste products are found in urine? _____

7. a) What is the main waste excreted by the skin? _____

   b) What other waste products does the skin excrete? _____ ,

      _____ , and _____

THE LUNGS AS ORGANS OF EXCRETION

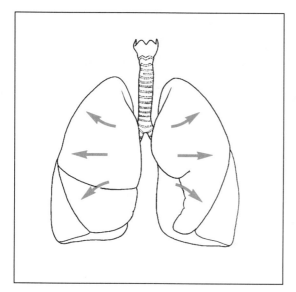

Figure B

The lungs excrete carbon dioxide.

What other wastes do the lungs excrete in small amounts?

_____

FIND YOUR KIDNEYS

Figure C shows where the kidneys are located.

Use your fists the same way to show where your kidneys are.

Figure C

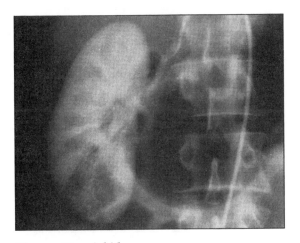

The main job of the kidneys is to filter out wastes from the blood. Inside each kidney, there are millions of tiny tubes. Many coiled capillaries are found in these tubes. As blood flows through the capillaries, water, salts, and urea are filtered out. These wastes leave the kidney and go into the kidney tube or **ureter** (yoo-REET-ur]. This liquid waste is called **urine** [YOOR-in]. Urine passes out of the ureter and collects in the bladder. Eventually, urine leaves the bladder through the **urethra** [yoo-REE-thruh].

Figure D *A kidney*

THE KIDNEY SYSTEM

Figure E shows the kidneys and the urinary tract. Remember, excretion is the removal of waste products from the body. Try to identify each part from its description. Write the letter of each part on the line next to the description.

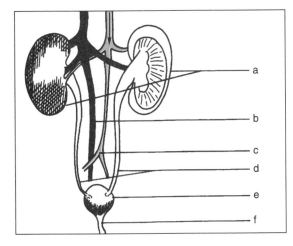

Figure E

1. The liquid waste formed in the kidneys is called

_____ .
urine, perspiration

2. How many kidneys does a person have? _____

_____ **3.** Kidneys—shaped like kidney beans. The kidneys make urine.

_____ **4.** Ureters—urine leaves the kidneys through these tubes. There are two ureters—one for each kidney.

_____ **5.** Bladder—a sac that collects and stores urine.

_____ **6.** Urethra—a tube that carries urine from the bladder to the outside of the body. There is one urethra.

_____ **7.** Blood vessels—carry blood to and from the kidneys.
(Hint: You will have two letters for this one.)

Figure F shows the parts of the skin.

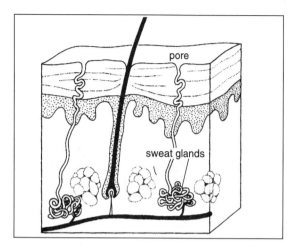

Figure F

1. The skin has many glands. One of these glands is shown in Figure F.

   a) What is this gland called?

   _____

   b) Name the liquid mixture made

   by this gland. _____

   _____

   c) List three materials that make up

   this mixture. _____ ,

   _____ , and _____

   d) The main waste excreted by the

   skin is _____ .

THE LIVER AS AN ORGAN OF EXCRETION

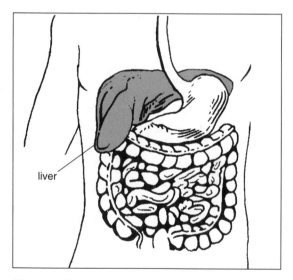

Figure G

The lungs, skin, and kidneys are the main organs of excretion. They pick up waste materials from the blood, and excrete these wastes directly.

The liver is an organ of excretion too. But the liver itself does not excrete wastes. Other organs do. That is why the liver is a secondary organ of excretion.

The liver handles cell wastes in several ways:

- The liver WEAKENS certain harmful substances. The substances become harmless.

- The liver CHANGES some harmful substances. They become useful substances.

For example: The liver makes bile from harmful substances. Bile is important in fat digestion. After the bile works on fat, it is eliminated from the bowels.

- The liver COMBINES certain harmful substances. It makes them ready for excretion.

For example: The liver combines two harmful wastes—ammonia and some carbon dioxide. Together, these substances form urea. The blood carries the urea to the kidneys. It becomes part of urine. Then, it is excreted.

ammonia + carbon dioxide → **urea**

poisonous waste

harmful waste   excreted by the kidneys

- The liver also BREAKS DOWN dead red blood cells. They pass into the digestive tract. Then, they are eliminated with the solid waste by the bowels.

YOUR OWN WORDS

Answer the following questions in brief sentences. Use your own words.

1. Why is the liver called a secondary organ of excretion? _____

   _____

   _____

2. Describe the four ways the liver handles cell wastes.

   a) _____

   b) _____

   c) _____

   d) _____

3. a) Bile is made from _____ substances.
   harmless, harmful

   b) What job does bile do? _____

c) What happens to bile after it does its work? _____

d) Which organ makes bile? _____

4. a) Ammonia and carbon dioxide combine to form _____ .

b) Which organ makes urea? _____

c) Urea is a _____ substance.
      helpful, harmful

d) Which organ excretes most of the body's urea? _____

5. a) Which organ breaks down red blood cells? _____

b) How are dead red blood cells eliminated from the body?

_____

6. The liver excretes wastes _____ .
      by itself, through other organs

REACHING OUT

Your body cells make more waste products when you are very active. What does your heart do to help eliminate the extra wastes?

_____

_____

_____

_____

_____

What are the sense organs?

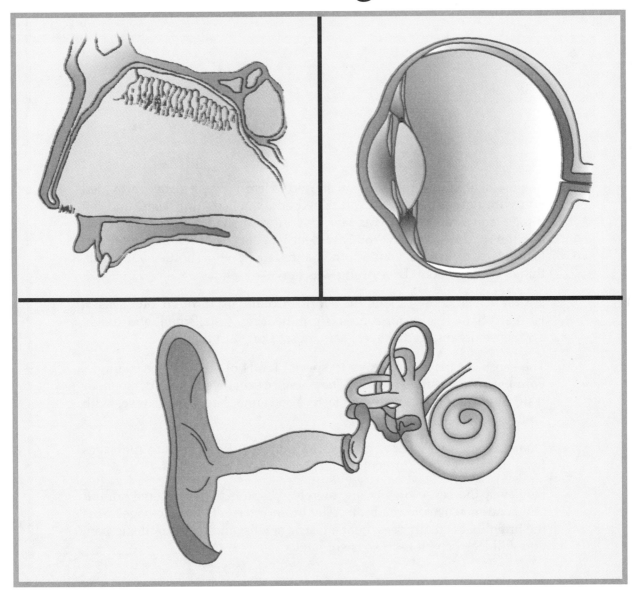

LESSON 19 | What are the sense organs?

The moment you were born, you started to learn. Every instant provided you with new experiences. You tasted and smelled. You heard and felt things. You saw. At first you saw only blurred shadows. Then, day-by-day the shadows cleared. You saw clearly . . . and you remembered! You learned to recognize your mother and father, your bottle, your crib, things in your room. Everything was new.

We learn and know about the world around us through our senses. Humans have five major senses: sight, hearing, taste, smell, and touch. The sense organs are the eyes, ears, nose, skin, and tongue.

The sense organs are sensitive to special kinds of stimuli. For example, your eyes are sensitive to light. They are not sensitive to sound, smell, or taste. Your tongue is sensitive to taste. You cannot hear, smell, or see with your tongue.

Our senses tell us what is going on around us. Responses to messages from our senses help protect us and keep us alive.

However, the sense organs are merely receptors. They receive stimuli and send messages to the brain. The brain interprets the messages. It is the brain that actually sees, hears, tastes, smells, and feels. In this lesson, you will learn about each sense organ.

THE EYES

Check with Figure A as you read about the different parts of the eye. The cornea [KOR-nee-uh] is the transparent "window" of the eye, where light enters the eye. The iris [Y-ris] is the circular colored part of the eye. The iris gives the eye its color. The pupil [PYOO-pil] is the opening in the middle of the iris that changes size, depending upon the amount of light that enters the eye. The lens helps focus light. The retina [RET-in-uh] is the light sensitive nerve layer of the eye. The optic nerve leads from the retina. It carries light messages to the brain. The brain then interprets the messages.

Identify the parts of the eye in Figure A. Write the correct letter on the lines provided.

_____ **1.** retina

_____ **2.** pupil

_____ **3.** optic nerve

_____ **4.** iris

_____ **5.** cornea

_____ **6.** lens

Figure A *The eye*

7. The optic nerve leads to the _____ .

THE EARS

Sound is caused by vibrations. The path of vibrations through the ear are described below. Identify each part by letter.

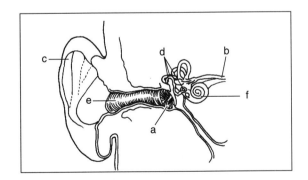

Figure B *The ear*

_____ **1.** Air vibrations are "gathered" by the outer ear.

_____ **2.** The vibrations pass through the ear canal.

_____ **3.** The vibrations cause the ear drum to vibrate.

_____ **4.** The ear drum passes on the vibrations to <u>three</u> <u>tiny</u> <u>ear</u> <u>bones</u>.

_____ **5.** The last bone passes on the vibrations to the snail-shaped <u>cochlea</u>.

_____ **6.** The cochlea is lined with microscopic hairs and contains a liquid. They vibrate. The vibrations pass to the <u>auditory</u> <u>nerve</u>.

7. To where does the auditory nerve lead? _____

THE TONGUE

The tongue is sensitive to certain chemicals. You know that there are different kinds of tastes—sweet, sour, bitter, and salty. The tongue has four kinds of specialized receptors called taste buds. Each kind of taste bud is sensitive to one kind of taste. They are located at different parts of the tongue.

□ sweet
▲ sour
● bitter
◗ salt

Figure C   *The tongue*

1. The taste buds at the rear of the tongue are sensitive to

_____ tastes.

2. The taste buds at the middle-sides

are sensitive to _____
tastes.

3. The taste buds at the very tip of the tongue are sensitive to _____ tastes.

4. The taste buds on the sides, but closer to the front, are sensitive to

_____ tastes.

5. Taste buds send messages to the _____ .

THE NOSE

Like the tongue, the nose is sensitive to certain chemicals. When you inhale through the nose, the air enters a space called the nasal cavity. The upper part of the nasal cavity contains many odor-sensitive nerve endings. The nerve endings "pick up" odors. All the nerve endings join as a single olfactory nerve.

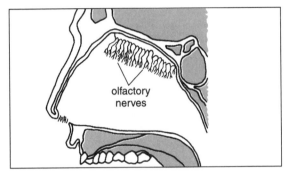

Figure D   *The nose*

1. Where do you think the olfactory

   nerve ends? _____ .

THE SKIN

There are several senses of "feel," pain, heat, cold, simple touch, and pressure. The skin has specialized receptors for each. Some areas of the skin are more sensitive than others.

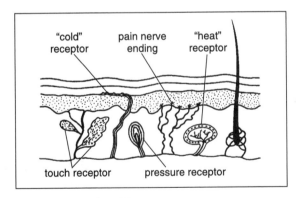

Figure E   *The skin*

1. More receptors result in

   _____ sensitivity.
   increased, decreased

2. What do "feel" receptors sense?

   _____

3. What interprets "feel" messages?

   _____

117

FILL IN THE BLANK

Complete each statement using a term or terms from the list below. Write your answers in the spaces provided. Some words may be used more than once.

taste	smell	iris	pain
sound	cochlea	sour	heat
touch	nose	bitter	cold
light	sweet	salty	pressure

1. The eyes are sensitive to _____ .

2. The ears are sensitive to _____ .

3. The nose is sensitive to _____ .

4. The tongue is sensitive to _____ .

5. The skin is sensitive to _____ .

6. The _____ is the colored part of the eye.

7. The _____ is the snail-shaped structure in the ear.

8. The four tastes the tongue is sensitive to are _____ ,

    _____ , _____ , and _____ .

9. The olfactory nerves are found in the _____ .

10. The skin can "feel" _____ , _____ , _____ ,

    _____ , and simple _____ .

WORD SCRAMBLE

Below are several scrambled words you have used in this Lesson. Unscramble the words and write your answers in the spaces provided.

1. ULISTSUM _____

2. SENO _____

3. CHUOT _____

4. GOTENU _____

5. PENSEROS _____

What is the nervous system?

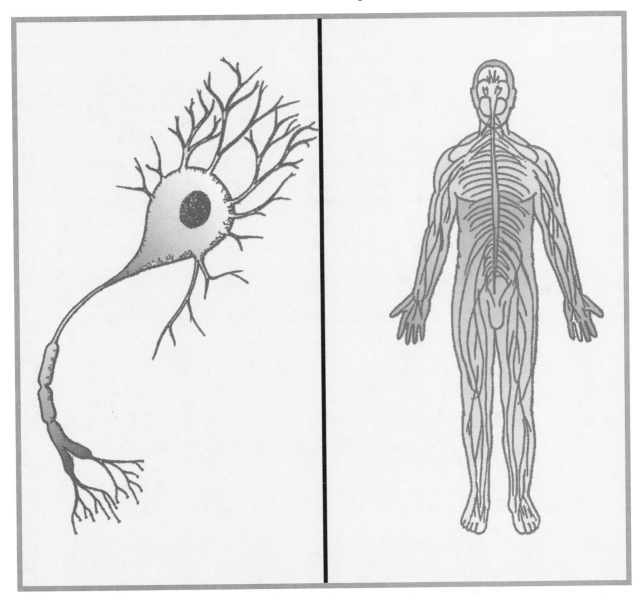

KEY TERMS

nervous system: body system made up of the brain, the spinal cord, and all the nerves that control body activities

neuron: nerve cell

LESSON 20 | What is the nervous system?

Every school has an office. It is a very important place. Messages come into the office. Messages go out. Most plans for the whole school are made in the office.

In your body, the job of receiving and sending messages is done by the **nervous system**. The nervous system controls all of your body's activities. The nervous system is made up of the brain, the spinal cord, and branching nerves.

The brain and spinal cord alone make up the central nervous system.

You have learned that the sense organs receive stimuli. But what happens to the stimuli after they are received? For example, how do you decide to answer the telephone, or raise your hand in class?

This is how the nervous system works:

• Stimuli from the sense organs change to electrical signals.

• These electrical signals do not stay in the sense organs. Nerves carry the signals to the brain and spinal cord.

• The brain decides what each stimulus is. The brain also decides how to respond to each stimulus.

• Nerves carry "what to do" messages away from the brain. The messages go to the part of the body that will answer or respond to the stimuli.

Most "what to do" messages go to muscles. Some, however, go to glands. Most responses are carried out by muscles.

Note: In some cases, the spinal cord, not the brain, receives and sends messages of how to respond to a stimulus. You will learn more about this in Lesson 22.

The nerves of the nervous system are made up of nerve cells. Each nerve cell is called a **neuron** [NOOR-ahn].

Neurons are well suited to performing their job of carrying messages. A group of neurons looks like a string of space-age telephones.

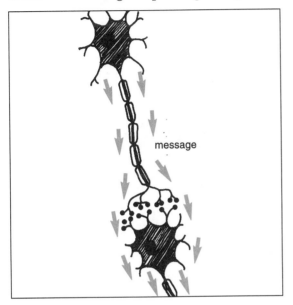

Look at Figure A. It shows a message moving along two neurons.

Neurons form a pathway along which electrical signals travel. At one end of the pathway is a sense organ. At the other end is the muscle or gland that responds to the stimuli.

Figure A

THE SPINAL CORD

Thirty-one pairs of nerves branch out from your spinal cord. These nerves are inside the spinal column (backbone). The backbone protects the nerves.

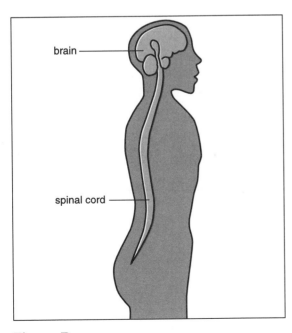

Your spinal cord runs down the center of your back. It extends from the base of the brain to the tailbone.

Some emergency responses must happen extra fast. There is no time for the brain to decide how to respond. Delay can cause severe injury—or even death.

In these cases, the spinal cord—not the brain, sets up the response. The response takes place even before the message reaches the brain.

These emergency responses to stimuli are called reflexes. You will learn more about reflexes in Lesson 22.

Figure B

- Some nerves carry messages to the brain and spinal cord.

- Other nerves carry messages away from the brain and spinal cord.

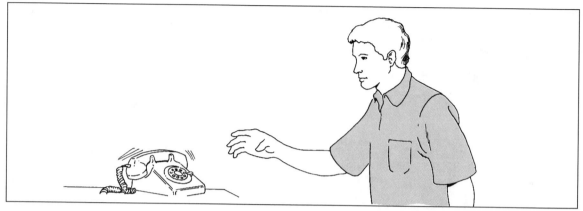

Figure C

1. Nerves that carry stimuli lead _____ the brain and spinal cord.
   to, away from

2. Nerves that carry messages for responses lead _____ the spinal cord.
   to, away from

3. In what energy form are nervous system signals? _____

FIND THE PARTS

Find the parts of the nervous system. Write their names on the correct lines. Choose from the parts listed below.

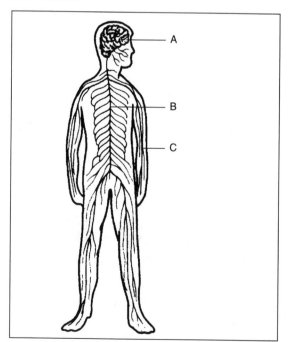

spinal cord
branching nerves
brain

Name the parts that make up the central nervous system.

A._____

B._____

C._____

Figure D

Complete each statement using a term or terms from the list below. Write your answers in the spaces provided. Some words may be used more than once.

nerves	sends	muscles
one direction	spinal cord	to
receives	away from	brain
backbone	stimuli	response

1. The nervous system _____ and _____ messages.

2. The parts of the nervous system are: the _____ , _____ ,

   and _____ .

3. Nerves carry messages in only _____ .

4. Some nerves carry messages _____ the brain and spinal cord. Some

   nerves carry messages _____ the brain and spinal cord.

5. _____ are carried to the brain and spinal cord by nerves.

6. Messages of _____ are carried away from the brain and spinal cord.

7. The _____ "decides" what to do about most stimuli.

8. Most messages of response are sent to _____ .

9. Most responses are carried out by _____ .

10. The spinal cord is protected by the _____ .

MATCHING

Match each term in Column A with its description in Column B. Write the correct letter in the space provided.

	Column A	**Column B**
_____	1. parts of the nervous system	a) an action
_____	2. stimulus	b) brain, spinal cord, and nerves
_____	3. response	c) a signal to do something
_____	4. brain and spinal cord	d) carry messages
_____	5. nerves	e) central nervous system

Figure E

1. Name the sense organs.

   _____ _____

   _____ _____

   _____

2. Four of the sense organs send nerves right into the brain. Which organs are they? (Think of your own body.)

   _____ _____

   _____ _____

3. Most nerves of one of the sense organs go to the spinal cord before they go to the brain. Which organ is this? (Think of your own body.)

   _____

REACHING OUT

Responses that are planned are voluntary responses. Responses that are not planned are involuntary responses.

1. What was the last voluntary response you did?

   _____

   _____

2. Can you name an involuntary response that you are probably doing right now?

   _____

   _____

Lesson **21**

What are the parts of the brain?

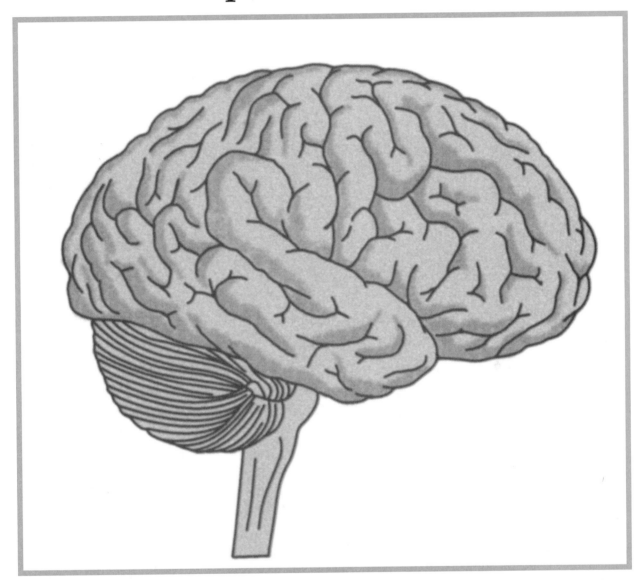

KEY TERMS

cerebrum: large part of the brain that controls the senses and thinking

cerebellum: part of the brain that controls balance and body motion

medulla: part of the brain that controls many vital functions, such as heartbeat and breathing

LESSON 21 | What are the parts of the brain?

The brain is the control center of your body. The brain is made up of a mass of nervous tissue. It is protected by your skull.

The main job of the brain is to receive messages and decide what to do. These messages may come from inside or outside your body. Your brain responds to the messages and then controls almost all of your body's activities.

The brain is made up of three main parts. They are the **cerebrum** [suh-REE-brum], the **cerebellum** [ser-uh-BELL-um], and the **medulla** [muh-dull-uh].

Different parts of the brain control different activities.

CEREBRUM   The cerebrum is the largest part of the brain. It controls the senses, thought, memory, and learning. It also controls certain voluntary muscles. You use voluntary muscles for activities like walking, talking, and writing.

CEREBELLUM   The cerebellum is located at the back of the brain. It works with the cerebrum to control voluntary muscles. The cerebellum controls body movements. The cerebellum also helps you keep your balance.

MEDULLA   The medulla is the smallest part of the brain. It is a thick stalk at the base of the skull. The medulla connects the brain to the spinal cord. It controls many vital involuntary functions. For example, the medulla controls breathing, digestion, and heartbeat. It also controls sneezing and blinking.

Look at Figure A. It shows the parts of the cerebrum that control certain activities. Then answer the questions below.

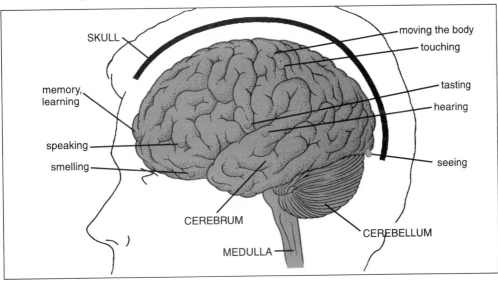

Figure A

1. What might happen if you were hit very hard on the back of your head? _____

    _____

2. What might happen if you were hit very hard on the front of your head? _____

    _____

3. What might happen if you were hit hard on the side of your head—towards the

    middle? _____

4. The brain is one of the most protected parts of your body.

    a) What protects your brain? _____

    b) Why does it protect so well? _____

    c) Of what is it made? _____

5. **a)** What is the largest part of the brain? _____

    b) What is the smallest part of the brain? _____

Label the main parts of the brain.

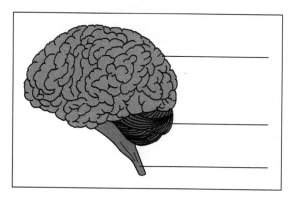

Figure B

COMPLETE THE CHART

Twelve actions are listed below. Each action is controlled by a different part of the brain. Place a check (✔) in the proper box (or boxes) for each action.

	ACTION	CONTROLLED BY		
		cerebrum	cerebellum	medulla
1.	hearing			
2.	seeing			
3.	moving the body			
4.	heartbeat			
5.	tasting			
6.	balance			
7.	sneezing			
8.	learning			
9.	breathing			
10.	speaking			
11.	memory			
12.	blinking			

MULTIPLE CHOICE

In the space provided, write the letter of the word that best completes each statement.

_____ 1. The control center of your body is your
  a) eye.   b) heart.
  c) brain.   d) lungs.

_____ 2. The largest part of the brain is the
  a) cerebrum.   b) cerebellum.
  c) medulla.   d) nerves.

_____ 3. The main job of the brain is to carry
  a) oxygen.   b) blood.
  c) messages.   d) hormones.

_____ 4. The brain is connected to the spinal cord by the
  a) cerebrum.   b) medulla.
  c) cerebellum.   d) inner ear.

_____ 5. Heartbeat and breathing are controlled by the
  a) cerebrum.   b) kidneys.
  c) cerebellum.   d) medulla.

MATCHING

Match each term in Column A with its description in Column B. Write the correct letter in the space provided.

Column A

_____ 1. medulla

_____ 2. cerebellum

_____ 3. skull

_____ 4. cerebrum

_____ 5. brain

Column B

a) controls learning

b) control center

c) smallest part of brain

d) protects brain

e) controls balance

SCIENCE *EXTRA*

The Eyes Have It

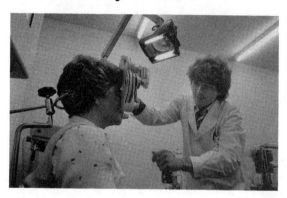

All body organs are subject to normal wear and tear as well as disease. The eyes are no exception.

An eye disorder that is very common among the elderly is called cataracts. A cataract affects the lens of the eye. The lens has two functions. It bends light that passes through it on its way to the retina (the nerve layer of the eye). The lens also changes its shape slightly to focus for close vision, such as reading and sewing.

A healthy lens is transparent. It is clear as glass. Sometimes, however, for reasons not fully understood, one or both lenses may become cloudy. A clouded lens is called a cataract. A cataract reduces vision greatly. It allows light to pass through. However, there is no detail. Only a glaring jumble of meaningless light reaches the retina. The vision may become so poor that a person with advanced cataracts in both eyes may be considered legally blind.

Fortunately, there is an answer to cataract problems. It is not possible to clear the cloudiness of the affected lens. However, it is possible to remove the lens. Surgeons have been doing this for centuries.

Cataract removal is only part of the cure for cataracts. Until about 20 years ago, most people who had undergone surgery had to wear very strong eyeglass lenses in order to see clearly. The lenses substituted for the power of the natural lens that had been removed. Vision usually was quite good, but not very natural. Everything was seen magnified; side vision was distorted, and it was difficult to judge distances. A long time was needed to "get used" to this new way of seeing. The eyeglass lenses also were very thick and heavy. They looked "like milk bottles," and "weighed a ton."

Fortunately, the latest chapter of the "cataract story" has been written and it is very good. Thanks to the combined efforts of chemists, biotechnicians, and eye surgeons, a lens substitute has been perfected. As soon as the lens with a cataract is removed, a plastic substitute, or implant is inserted. The implant allows for very natural vision. And in many cases, the patient needs no eyeglasses, or only a pair for general distance use. Usually, "average" power eyeglasses are needed for close vision.

What is a reflex?

KEY TERM

reflex: automatic response to a stimulus

LESSON 22 | What is a reflex?

The moment you are born you do certain things by yourself. You cry, you yawn, your eyes blink, your lips search for food.

You were not taught how to do these things. You were born knowing how to do them.

These kinds of responses are called **reflexes**.

There are many kinds of reflexes. But they are all alike in certain ways.

• Reflexes are not learned. They are <u>inborn</u>.

• You do not control or think about reflexes. They happen by themselves. They are automatic and involuntary responses.

• A reflex is done exactly the same way every time.

In most cases, you do not know a reflex is happening. For example, jumping away from a speeding car is a reflex. You respond without thinking. You know about it only after the response has taken place.

The same is true when you touch a hot pot. You pull your hand away <u>before</u> your brain "feels" the pain.

Reflexes are very important. They protect us and help us stay alive. Reflexes control most of our body organs.

Most reflexes occur very quickly. This is because reflexes do not involve the brain. They are controlled by the spinal cord. Look at the example below to help you understand the path of a reflex.

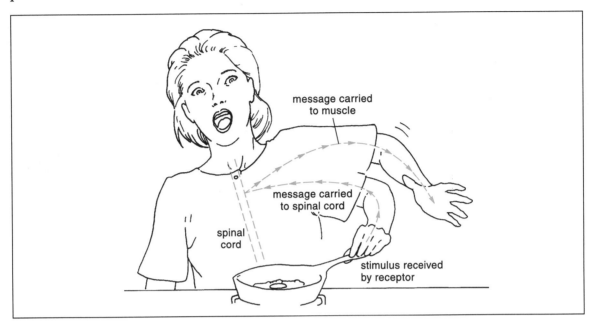

Figure A

The <u>stimulus</u>: touching a hot object.

The <u>response</u>: pulling the hand away.

- FIRST   Cells in the skin detect heat. Nerves send the message of "heat" to the spinal cord. The spinal cord decides what to do.

- SECOND   Nerves carry this message of "what to do" away from the spinal cord. It goes to the muscles of the hand.

- THIRD   The message tells the muscles to "let go" of the hot object.

At this point, the brain does not know what is happening. However, while messages are moving along the reflex path, the spinal cord is sending messages to the brain. Once the brain receives these messages, it sends messages to your hand. Then, you feel pain. That is why a reflex action is usually followed by a loud "OUCH!"

Reflexes control important body organs.

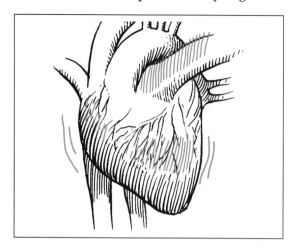

Figure B   *Reflexes control your heartbeat.*

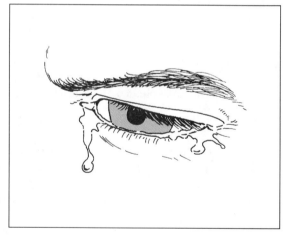

Figure C   *Reflexes control your breathing.*

1. What happens to your heartbeat if you are excited?

   _____

2. What happens to your heartbeat if you are asleep?

   _____

Reflexes protect you from injury.

Figure D   *When you trip, your hands move automatically to protect your face.*

Figure E   *When dust gets in your eyes you tear, and your eyelids flutter—automatically.*

3. What part of your body do you seem to protect first automatically?

   _____

4. How does this reflex protect you?

   _____

COMPLETING SENTENCES

Choose the correct word or term for each statement. Write your choice in the spaces provided.

1. Reflexes are _____ .

learned, inborn

2. You _____ control reflexes.

can, cannot

3. Reflexes _____ planned.

are, are not

4. Reflexes _____ happen by themselves.

do, do not

5. You _____ know that most reflexes are happening.

do, do not

6. Reflex responses are carried out by _____ . (Careful, this one is tricky.)

nerves, muscles

7. Most reflexes are very _____ .

slow, fast

8. A reflex always happens _____ .

the same way, in different ways

9. Reading _____ a reflex.

is, is not

10. Blinking when something enters your eye _____ a reflex.

is, is not

WORD SCRAMBLE

Below are several scrambled words you have used in this Lesson. Unscramble the words and write your answers in the spaces provided.

1. PONSESER   _____

2. XERFEL   _____

3. NAIP   _____

4. NORBNI   _____

5. SULUMITS   _____

An instinct [in-STINKT] is like a reflex. It is inborn and automatic. And it happens the same way, every time. BUT, an instinct is much more complex than a reflex.

There are many instincts. Animals depend on instincts more than humans. For example, a bird uses instinct to build its nest. A bird can build a nest even if it has never seen one built.

Nest-building is complicated. A bird must choose a nesting place. It must select nesting materials. Then it must put the nest together.

Scientists believe that an instinct is a series, or chain, of reflex responses. Each response leads to another. And, if one response in the "chain" is not done, the "instinct" will not be completed—or it will not be completed correctly.

For example, if a bird cannot put its nesting materials together correctly, the nest will not be built. Or if built, the nest will not be a good one.

Now, complete these sentences about instincts.

1. Instincts are _____ .
   learned, inborn

2. Both reflexes and instincts are _____ .
   thought out, automatic

3. Instincts are _____ complex than reflexes.
   more, less

4. An instinct is a series of inborn _____ .
   stimuli, responses

5. For an instinct action to be completed, all steps leading to the action must be

   _____ .

REACHING OUT

In what kind of jobs are very fast reflexes especially important?

_____

_____

Lesson **23**

What is the endocrine system?

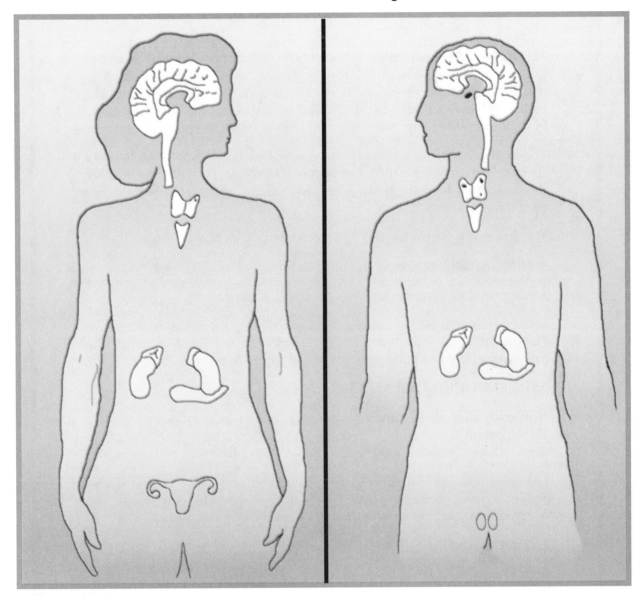

KEY TERMS

endocrine system: body system made up of about ten endocrine glands that help the body respond to changes in the environment

hormone: chemical messengers that regulate body functions

LESSON 23 | What is the endocrine system?

Conditions inside and outside your body are always changing. Some of these changes can be harmful. Your body has two organ systems to help it adjust to these changes. They are the nervous system and the **endocrine** [EN-duh-krun] **system**. You learned about the nervous system in the last few lessons. Now you will learn about the endocrine system.

The endocrine system is made up of many glands. These glands make chemical messengers called **hormones** [HAWR-mohns]. Hormones are chemicals that help the body adjust to changes. But that is not all they do. Hormones also:

- help control chemical reactions in the body,

- affect maturity and reproduction, and

- help regulate physical and mental development.

You know about some glands, like the salivary glands and the sweat glands. But these glands are not endocrine glands. What, then, are the differences between endocrine glands and non-endocrine glands?

NON-ENDOCRINE GLANDS

Non-endocrine glands are also called **exocrine** [EK-suh-krun] glands. The substances made by exocrine glands flow through tubes called ducts. These chemicals empty directly into the place they will be used.

For example, the salivary glands secrete digestive enzymes. These enzymes flow through ducts. The ducts empty directly where the enzymes are used. The saliva empties directly into the mouth.

ENDOCRINE GLANDS

Endocrine glands are different. Endocrine hormones

- do not move through ducts, and

- they do not empty directly where they will be used.

Hormones from the endocrine glands empty into the bloodstream. The blood then carries the hormones to the places where they do their work.

The glands that make up the endocrine system are shown in Figure A. Each gland is described below. Try to identify each gland from its description.

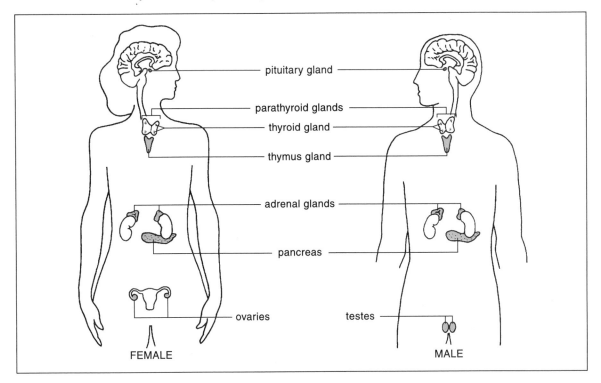

Figure A

PITUITARY [pi-TOO-uh-ter-ee] **GLAND**—small round gland at the base of the brain.

THYROID [THY-royd] **GLAND**—butterfly-shaped gland at the base of the neck.

PARATHYROID GLANDS—four small glands embedded in the back of the thyroid gland.

THYMUS—located in the upper chest.

ADRENAL [uh-DREE-nul] **GLANDS**—two separate glands. One adrenal gland is located on the top of each kidney.

ISLETS OF LANGERHANS—glands scattered throughout the pancreas.

GONADS—(sex glands) These glands are different in males and females.

In males, the gonads are called TESTES. There are two testes. They are located in the lower groin.

In females, the gonads are called OVARIES. There are two ovaries. The ovaries are almond-shaped. They are located in the lower abdomen area.

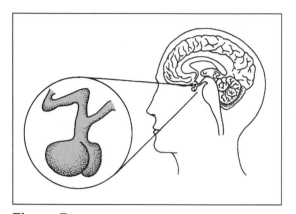

Figure B

PITUITARY GLAND. This gland makes many hormones. Some regulate growth and the production of sex cells.

The pituitary gland also controls other glands. For this reason, it is called the "master gland."

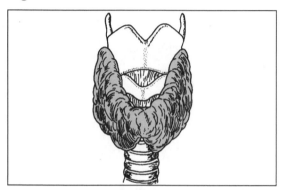

Figure C

THYROID GLAND. This gland regulates metabolism. Metabolism is all of the chemical reactions that take place in an organism.

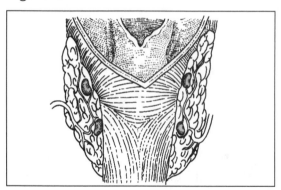

Figure D

PARATHYROID GLANDS. Regulate use of the minerals calcium and phosphorus.

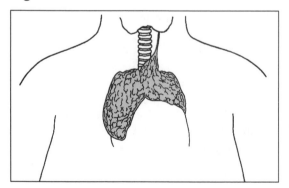

Figure E

THYMUS. Controls the growth of certain white blood cells that help the body fight infection.

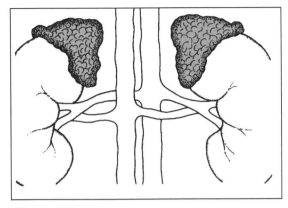

Figure F

ADRENAL GLANDS. Controls muscle reaction in times of stress—especially sudden stress.

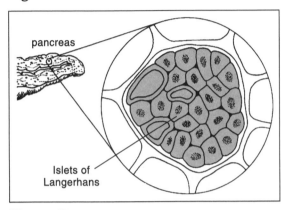

Figure G

ISLETS OF LANGERHANS. Produce the hormone insulin. Insulin helps control the amount of sugar in the blood.

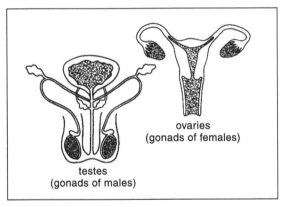

Figure H

GONADS. The gonads are the sex glands. The sex glands are different in males and females. Sex glands produce the cells needed for reproduction.

The sex glands of males are the testes. The testes make male reproductive cells called sperm. The testes also produce male sex hormones.

- Male sex hormones control the development of adult male characteristics—such as a deeper voice and facial hair.

The sex glands of females are the ovaries. The ovaries make female reproductive cells called eggs. They also make female sex hormones.

- Female sex hormones control the development of adult female characteristics—such as widening of the hips.

FILL IN THE BLANK

Complete each statement using a term or terms from the list below. Write your answers in the spaces provided.

changing	hormones	endocrine
ovaries	adjusting	chemical
all parts	harmful	bloodstream
ducts	nervous	testes

1. The body's environment is always _____ .

2. Some of these changes can be _____ .

3. The body is always _____ to these changes.

4. The two organ systems that help the body adjust to changes are the

   _____ system and the _____ system.

5. Endocrine glands make chemicals called _____ .

6. Endocrine glands do not have _____ . Their hormones empty into the

   _____ .

7. Blood carries hormones to _____ of the body.

8. Hormones help the body adjust to changes. They also help regulate

   _____ reactions in the body.

9. The male gonads are the _____ .

10. The female gonads are the _____ .

REACHING OUT

A person without a hat is said to be hatless. Some birds do not have properly developed wings. These birds are said to be wingless or flightless.

Endocrine glands are called by another name also. What is that name? Use the above examples and what you know about endocrine glands to find the answer.

(Hint: The name ends in "less.") _____

What is behavior?

KEY TERMS

behavior: response of an organism to its environment

conditioned response: behavior where one stimulus takes the place of another

LESSON 24 | What is behavior?

At home or in school, behavior means being good or bad. To a scientist, **behavior** means all kinds of actions. It means how we react to all stimuli. It also means how we learn.

Scientists study behavior. They try to find out why we behave the way we do. They also try to find out how behavior can be changed.

One of the scientists who studied behavior was a Russian named Ivan Pavlov. About 75 years ago, Pavlov experimented with dogs in a special way. He knew that dogs drool every time they see or smell food. This is a normal reflex response.

Pavlov wanted to find out if this reflex could be changed. This is what he did: Pavlov rang a bell every time he brought food to a dog. Each time, the dog drooled. Remember, food always makes a dog drool.

Then Pavlov did something new. He just rang the bell. He did not give the dog any food. What do you think happened? The dog's mouth watered even though no food was there.

The dog had learned that the bell always meant food. The bell had now taken the place of the food as a stimulus that causes drooling.

A behavior where one stimulus takes the place of another stimulus is called a **conditioned** [kun-DISH-und] **response**. Pavlov's dog was conditioned to respond to a new kind of stimulus. What was this new stimulus? Conditioning is a very simple form of learning.

TRACING PAVLOV'S EXPERIMENT

Figure A

Figure B

Figure C

Look at Figure A.

1. What stimulus is reaching the dog?

   _____

2. What is the response to this

   stimulus? _____

3. The dog _____ control
   _{can, cannot}

   this response.

4. The response is _____ .
   _{inborn, learned}

Look at Figure B.

5. What new stimulus has been added?

   _____

6. Does this stimulus usually cause

   drooling? _____

7. Both stimuli reach the dog

   _____ .

   at the same time, at different times

Look at Figure C.

8. Which stimulus has been taken

   away? _____

9. Which stimulus is reaching the dog?

   _____

10. The ringing bell alone now

    _____ make the dog drool.
    does, does not

11. Without conditioning, a ringing bell

    _____ cause drooling.
    does, does not

12. The dog drools because the sound of the bell has been connected to

_____ .

13. The dog has experienced a simple kind of learning called _____ .

14. This kind of behavior is called a _____ .
reflex, conditioned response

Now look back at Figure B.

15. The diagram does not show it, but we know that both stimuli reached the dog

together _____ .
only a few times, many times

MORE ON CONDITIONING

Figure D

The dog has learned to connect the sound of "sit up" with a treat.

Being given treats has conditioned him to sit up.

Figure E

This child has learned to get attention by crying.

Figure F

What two stimuli have these fish learned to connect?

_____

146

TRUE OR FALSE

In the space provided, write "true" if the sentence is true. Write "false" if the sentence is false.

_____ **1.** Behavior means only being good or bad.

_____ **2.** Behavior means all kinds of responses.

_____ **3.** Scientists study how living things behave.

_____ **4.** Behavior is always the same.

_____ **5.** Pavlov studied apes.

_____ **6.** We can learn about people, by studying animals.

_____ **7.** A reflex is learned.

_____ **8.** A conditioned response is a learned response.

_____ **9.** A bell always causes a dog to drool.

_____ **10.** A bell can never cause a dog to drool.

MATCHING

Match each term in Column A with its description in Column B. Write the correct letter in the space provided.

Column A

_____ **1.** Pavlov

_____ **2.** conditioning

_____ **3.** reflex

_____ **4.** behavior

_____ **5.** conditioned response

Column B

a) a simple form of learning

b) all kinds of responses

c) inborn response

d) studied dog behavior

e) a response with a changed stimulus

WORD SCRAMBLE

Below are several scrambled words you have used in this Lesson. Unscramble the words and write your answers in the spaces provided.

1. IORHEBAV _____

2. FLERXE _____

3. MULSSTIU _____

4. VVALOP _____

5. PONSERES _____

REACHING OUT

How can a conditioned response be unlearned? Use Pavlov's "drooling" experiment as an example.

_____

_____

_____

_____

_____

How do you learn?

KEY TERM

habit: learned behavior that has become automatic

LESSON 25 | How do you learn?

Do you remember when you were learning to ride a bike? Chances are it was not too easy. How many "flops" did you take? At first you thought about everything you had to do. You thought about the pedals. You thought about the steering. You do not think of these things now. You just ride off. After much practice, you have learned to ride a bike. You do it without thinking . . . you do it automatically.

Learned behavior that has become automatic is called a **habit**.

Many things you do every day are habits. Walking into a classroom each day and taking the same seat can become a habit. Actions that you practice a lot can become habits. Throwing a ball the same way over and over again may form a habit that improves your game.

You now know that learning can take place through conditioning and habit. But there are other ways. You can also learn through trial and error, memory, and reason.

- Trial and error means making mistakes and learning from them. You try different ways of doing something until you find the right way.

How would you find out which key in a batch fits a padlock?

- Memory keeps information stored in your brain. You can use this information anytime you need it.

What is your address? What are the names of your teachers? Your memory quickly gives you the answers to these questions.

- Reason helps you think out a problem carefully. You think of everything that might help you find the answer.

How would you go about trying to find a lost book? The first thing you probably would say is, "Now, let me see, where was I?" This is using reason.

Each of the figures below suggests a kind of learning. On the line under each figure, write the main kind of learning you think it is.

Choose from the following: **conditioning, habit, trial and error, memory, reason.**

Figure A

1. _____

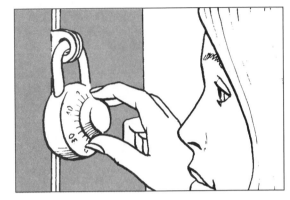

Figure B

2. _____

Figure C

3. _____

Figure D

4. _____

Figure E

5. _____

A SCHOOL FOR MICE?

Many animals, like mice for example, learn through trial and error. Some animals have good memories. Chimps, monkeys, and dolphins can even reason—but only in simple ways.

The mice in Figures F and G are going through a maze. The mazes are the same. There is a reward (a piece of cheese), at the end of maze G. There is no reward at the end of maze F.

Study Figures F and G. Think about them. Then answer the questions below.

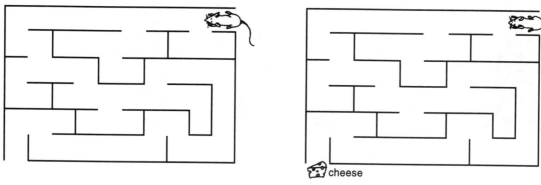

Figure F   **Figure G**

1. The mice will learn how to go through the maze by using

   _____ .
   trial and error, memory, reason

2. After they learn the maze, the mice will find their way by using

   _____ .
   trial and error, memory, reason

3. **a)** Which mouse will probably learn the maze faster? _____

   b) This shows that a reward _____ help learning.
   does, does not

MATCHING

Match each term in Column A with its description in Column B. Write the correct letter in the space provided.

Column A	Column B
_____ 1. trial and error	**a)** highest form of learning
_____ 2. habit	**b)** uses stored information
_____ 3. memory	**c)** uses changed stimulus
_____ 4. reasoning	**d)** learning from mistakes
_____ 5. conditioning	**e)** learning by practicing

FILL IN THE BLANK

Complete each statement using a term or terms from the list below. Write your answers in the spaces provided. Some words may be used more than once.

brain	memory	habit
trial and error	conditioning	reason
humans	nervous	

1. Five ways of learning are: _____ , _____ ,

   _____ , _____ , and _____ .

2. Learning by changing a stimulus is called _____ .

3. Learning by practice is called _____ .

4. Learning by trying different ways is called _____ .

5. Learning by storing knowledge is called _____ .

6. Learning by thinking is called _____ .

7. _____ use reason to solve problems.

8. Memory is information stored in the _____ .

9. The brain is part of the _____ system.

FUN WITH LEARNING: TRIAL AND ERROR

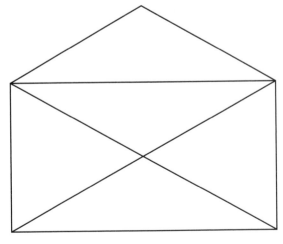

Figure H

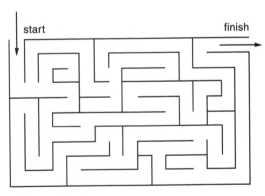

Figure I

1. Draw this figure without lifting your pencil.

2. Find your way through this maze.

Use the clues to complete the crossword puzzle. Highlight the words used in the lesson.

Clues

Across

4. An inborn, automatic act

5. Scientist who experimented with behavior

6. Bull is to cow as rooster is to

   _____

7. Good _____ gold

8. Go up

10. Not out

12. Part of the nervous system

13. Also

Down

1. They send and receive messages

2. All kinds of acts

3. This flows in body tubes

6. Learned automatic act

9. Tiny insect

11. Opposite of yes

What are the reproductive organs?

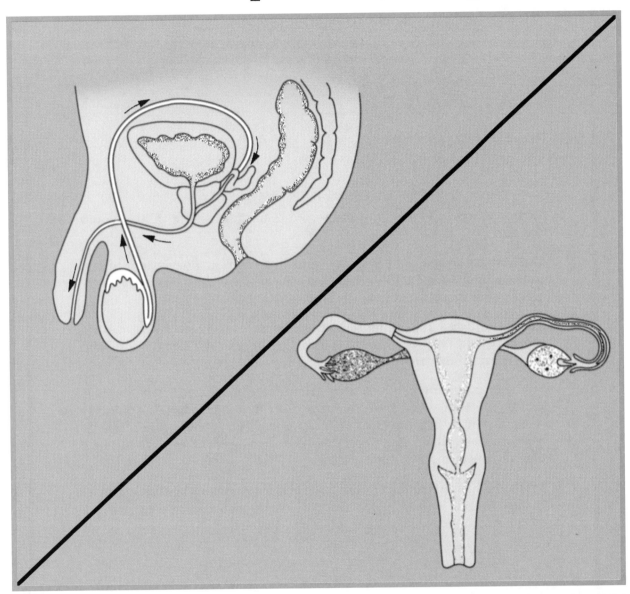

KEY TERMS

sperm: male reproductive cell

egg: female reproductive cell

ovaries: female reproductive organs

testes: male reproductive organs

LESSON 26 | What are the reproductive organs?

Reproduction is a vital life process. Without it, all living things would die out.

Unlike most of the other body systems, the reproductive system differs in males and females. These differences begin to show up as early as six weeks after a baby begins to develop.

There are two kinds of reproduction—asexual and sexual. In asexual reproduction, only one parent is needed. In sexual reproduction, two parents, one male and one female, are needed.

Humans and many plants and animals reproduce sexually. The method of reproduction varies from one organism to another, but one thing is certain. A male reproductive cell, a **sperm**, must unite with a female reproductive cell, an **egg**. Only then, can development and growth of a new organism begin.

What are the parts of the human male and female reproductive systems? In this lesson, we will study both the male and female reproductive systems, and what makes up each organ system.

The female reproductive system has four main parts. They are the **ovaries** [OH-vuhr-eez], the **oviducts** [OH-vuh-dukt], the **uterus** [YU-tur-us] and the **vagina** [vuh-JY-nuh]. Look at Figure A and find each organ as its function is explained.

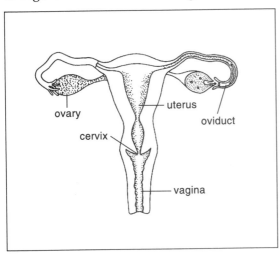

Figure A

The Ovaries

There are two ovaries, one on each side of the uterus. Each ovary is about the size and shape of a flattened, lumpy olive. The ovaries are the main female sex organs. The ovaries contain two kinds of cells. One kind of cell produces eggs. The other kind of cell produces hormones. These hormones are responsible for the development of secondary sex characteristics, (such as the growth of body hair, the growth of breasts) and the onset of puberty—the maturing of the reproductive system.

The Oviduct

There are two oviducts. Each oviduct extends from the uterus to one of the ovaries. The side of the oviduct closest to the ovary has fingerlike projections. Oviducts also are called fallopian [fuh-LOH-pee-un] tubes.

Once a month, an egg released by one of the ovaries, enters the oviduct. The egg moves through the oviduct and enters the uterus. Fertilization, when it takes place, occurs within one of the oviducts.

The Uterus

The uterus, or womb, is shaped somewhat like an inverted pear. It is a hollow thick-walled muscular organ. It is within the uterus that an embryo develops. The lower end of the uterus is called the cervix.

The Vagina

The cervix connects the uterus to the vagina. A baby moves through the vagina as it is being born. For this reason, the vagina also is called the birth canal.

Figure B shows the human female reproductive organs. Without looking back to the previous page, see if you can identify them by letter.

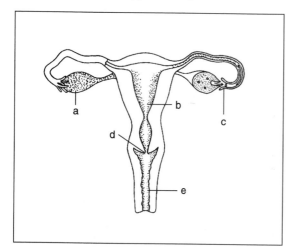

Figure B

1. uterus _____

2. ovaries _____

3. vagina _____

4. oviduct _____

5. cervix _____

Answer the following questions about the female reproductive system.

1. What are the female reproductive cells called? _____

2. Where are the eggs stored? _____

3. Where does an embryo develop? _____

4. Where does fertilization take place? _____

5. How does an egg reach the uterus? _____

MORE ABOUT EGGS

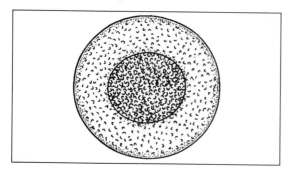

Figure C

A human egg is also is called an ovum. It is about the size of a pinpoint. This is large compared to other cells. In fact, an egg can be seen without a microscope.

Each egg has the potential to develop into an embryo, if it joins with a male sex cell.

At birth, a baby girl has all the egg cells she will have in her lifetime. However, the eggs are not fully mature. Egg cells begin to mature at the onset of **puberty** [PYOO-bur-tee]. Girls usually reach puberty between the ages of 10 and 14. Puberty is marked by the beginning of **menstruation** [men-stroo-WAY-shun].

The menstrual cycle occurs every 28–32 days. It is started by the release of hormones in the body.

- During the first stage, a hormone causes an egg to mature. The uterine wall begins to thicken with blood vessels.

- During the second stage, the egg is released from the ovary into the oviduct. This is called ovulation [oh-vyuh-LAY-shun].

- During the third stage, the uterine wall continues to thicken. This prepares the uterus for an embryo, if the egg has been fertilized.

- The fourth stage only occurs if the egg has not been fertilized. The tissue, blood, and mucus that were lining the uterine wall, break down and leave the body. This process is called menstruation.

The male reproductive system also has four main parts. They are the **testes** [TES-teez], the **urethra** [yoo-REETH-ruh], the **vas deferens** [DEF-uh-renz] and the penis. Look at Figure D and find each organ as its function is explained.

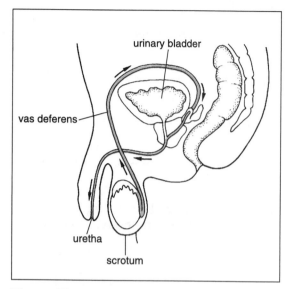

Figure D

The Testes

The testes are the main male reproductive organs. The singular form of testes is testicles. The testes, like the ovaries, contain two kinds of cells. One kind of cell produces sperm. The other kind of cell produces hormones, which are responsible for the development of secondary sex characteristics. The testes are located in a sac called the scrotum [SKROH-tum].

The Vas Deferens

The vas deferens is a tube that leads from each testicle into the urethra. When sperm are released they enter and move through the vas deferens into the urethra.

The Urethra

The urethra is a tube located inside the penis. As the sperm enter the urethra, several glands secrete fluid. The fluid helps the sperm move easier. The combination of the fluid and sperm is called semen. Semen is released through the penis during ejaculation.

The urethra also is part of the male's excretory system. Urine travels from the bladder out of the body through the urethra.

Answer the following questions about the male reproductive system.

1. What are the male reproductive cells called? _____

2. Where are sperm produced? _____

3. Into which tubes are sperm first released? _____

4. Name the tube through which sperm finally leave the body.

_____

A sperm has two parts, a head and a tail. Sperm are much smaller than eggs. You need a microscope to see them.

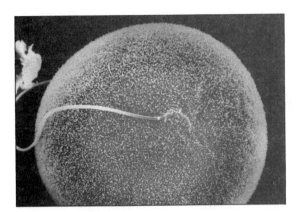

Figure E shows a single egg cell and several sperm cells. Notice how much larger the egg is.

Figure E

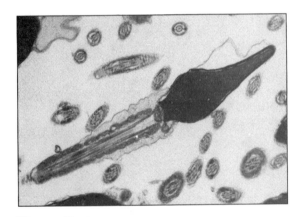

Figure F shows a magnified picture of sperm. Notice the head and the tail of the sperm.

Figure F

1. Of what use do you think the sperm's tail is? _____

_____

2. Does an egg have a tail? _____

MATCHING

Match each term in Column A with its description in Column B. Write the correct letter in the space provided.

Column A

_____ 1. cervix

_____ 2. eggs

_____ 3. ovaries

_____ 4. oviduct

_____ 5. scrotum

_____ 6. sperm

_____ 7. testes

_____ 8. urethra

_____ 9. uterus

_____ 10. vagina

_____ 11. vas deferens

Column B

a) tube that leads from testes to urethra

b) narrow end of the uterus

c) pocket of skin that holds the testes

d) organ in which an embryo develops

e) tube that carries sperm and urine to the outside of the body

f) long tube between the ovary and the uterus

g) main organs of the male reproductive system

h) female sex cells

i) male sex cells

j) organs that produce female sex cells

k) birth canal

REACHING OUT

The uterus is very muscular. Why do you think this is important?

_____

_____

_____

_____

Lesson 27

How does fertilization take place?

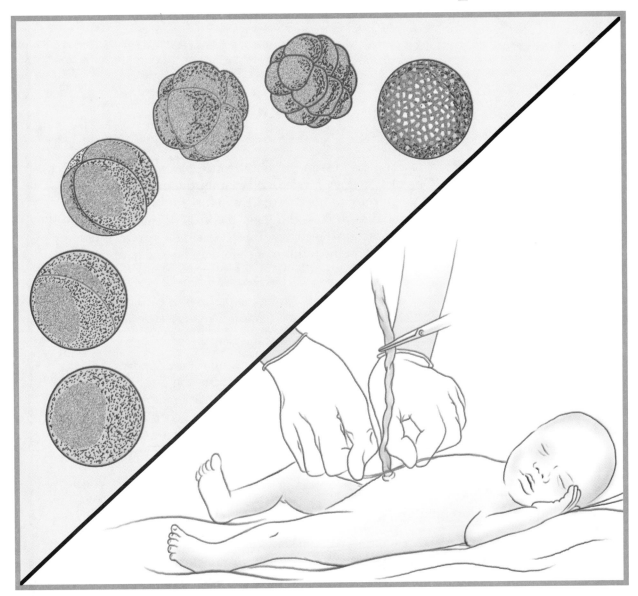

KEY TERMS

fertilization: joining of one sperm cell and one egg cell

zygote: fertilized egg produced by fertilization

embryo: hollow ball of cells formed by cell division of the zygote

placenta: organ through which an embryo receives nourishment and gets rid of wastes

LESSON 27 | How does fertilization take place?

As you have just learned in Lesson 26, about once every 28 days, a female ovulates. One of her eggs is released from an ovary and travels into the oviduct on its journey to the uterus. Tiny hairs line the oviduct. The motion of these hairs moves the egg towards the uterus. If **fertilization** [fur-tul-i-ZAY-shun] is to take place, a single sperm must reach and penetrate that egg in the oviduct. Fertilization is the joining of a sperm and an egg.

How do the egg and sperm meet up? An egg is passive. It cannot move by itself. A sperm cell, however, can move. The sperm has a long tail. The motion of this tail moves the sperm.

The fertilization race is on! During ejaculation, millions of sperm cells are released. However only one sperm can penetrate the egg. The egg gives off a chemical that attracts the sperm. The millions of sperm swim toward the egg. Only a small percent of the sperm reach the oviduct. Even fewer reach the egg. The sperm that reach the egg surround it. Each sperm tries to penetrate the egg. But there can be only one winner!

The successful sperm penetrates the egg and its nucleus. At that moment, fertilization takes place. The egg develops a protective membrane around itself to keep other sperm from penetrating the egg.

The fertilized egg, or **zygote** [ZY-gote], starts to divide. It moves into the uterus and attaches itself to the uterine wall. Cell division then proceeds at a faster rate. In nine months, a new person will be born.

Figures A, B, and C show the steps of human fertilization.

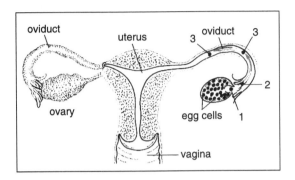

Figure A

1. An ovary releases a mature egg.

2. The egg enters the oviduct.

3. The egg is moved along the oviduct.

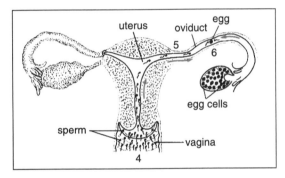

Figure B

4. Millions of sperm enter the female reproductive system through the vagina.

5. The sperm swim toward the oviduct and the egg.

6. Only a few sperm reach the egg.

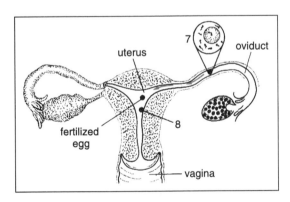

Figure C

7. One sperm penetrates the egg and its nucleus. Fertilization takes place.

8. The fertilized egg (zygote) attaches itself to the uterine wall. The fetus develops within the uterus.

1. How many eggs are usually released during ovulation? _____

2. Where does fertilization take place? _____

3. How many sperm fertilize an egg? _____

4. How do sperm move? _____

5. What is a fertilized egg called? _____

After fertilization, the zygote is a <u>single</u> cell. This is one case where one plus one equals one. The zygote divides by mitosis. Two attached cells are formed. Again they divide by mitosis and become four attached cells. The cell division continues until a hollow ball of cells is formed. The hollow ball of cells attaches itself to the uterine wall. The mass of cells is now called an **embryo** [EM-bree-oh]. All of the tissues and organs of the body form from the cells in the embryo.

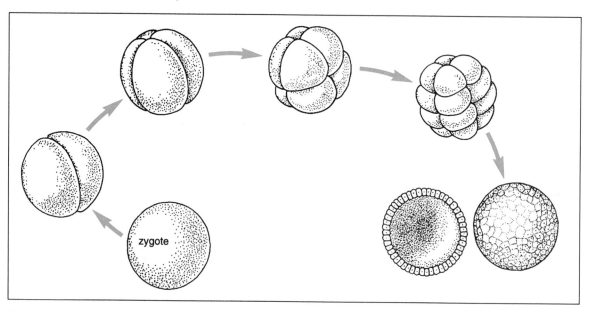

Figure D *Mitosis in a zygote*

The tissue that surrounds the embryo develops into a thick, flat structure called the **placenta** [pluh-SEN-tuh]. The embryo is attached to the placenta by the **umbilical** [un-BIL-ih-kul] cord. The umbilical cord is a ropelike structure that has two large blood vessels. One blood vessel carries nutrient-rich blood to the embryo. The other blood vessel carries wastes away from the embryo.

The developing embryo is also protected by a clear, fluid-filled sac. This sac is called the **amnion** [AM-nee-on].

1. What is an embryo? _____

   _____

2. How does an embryo receive nourishment? _____

   _____

   _____

3. Why do you think it is important for the embryo to get nourishment and to get rid

   of wastes? _____

   _____

Figure E

After about eight weeks, the embryo begins to develop a heart, brain, and nerve cord. The eyes and ears also begin to form. Arm and leg buds begin to form with fingers and toes. The cartilage in the embryo's skeleton starts to be replaced with bone. After bone replacement is complete, the embryo is called a fetus. The fetus looks more like an infant. The fetus continues to develop rapidly.

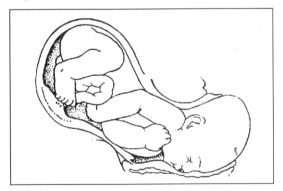

Figure F

After about nine months, the fetus is fully developed. It is ready! Hormones in the mother cause the uterus to tighten, or contract.

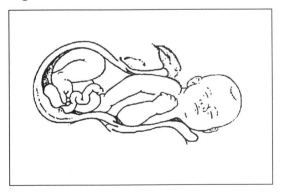

Figure G

The contractions of the uterus are called labor. The contractions become stronger and more frequent.

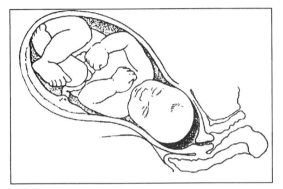

Figure H

The muscles in the uterine wall start to push the baby out.

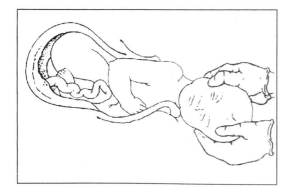

Figure I

Labor continues until the baby's body is pushed out of the mother's body. Further contractions push the placenta out of the mother's body.

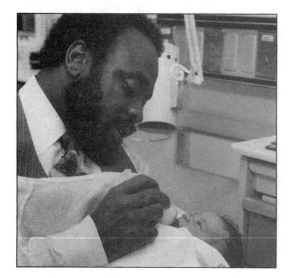

Figure J

The baby is born. Happy Birthday!

MATCHING

Match each term in Column A with its description in Column B. Write the correct letter in the space provided.

Column A	Column B
_____ **1.** amnion	**a)** ball of cells formed by cell division
_____ **2.** embryo	**b)** fertilized egg
_____ **3.** fetus	**c)** ropelike structure
_____ **4.** ovulation	**d)** embryo in which bone replacement is complete
_____ **5.** umbilical cord	**e)** clear fluid-filled sac
_____ **6.** zygote	**f)** release of a mature egg from an ovary

MULTIPLE CHOICE

In the space provided, write the letter of the word or phrase that best completes each statement.

_____ 1. A mature egg leaves an ovary and travels to an oviduct during
   a) menstruation.   **b)** ovulation.   **c)** mitosis.   **d)** fertilization.

_____ 2. The umbilical cord connects the embryo to the
   a) amnion.   **b)** cervix.   **c)** uterus.   **d)** placenta.

_____ 3. In males, both urine and sperm leave the body through the
   a) urethra.   **b)** scrotum.   **c)** gamete.   **d)** testes.

_____ 4. A developing embryo is cushioned and protected by
   a) an umbilical cord.   **b)** the ovaries.   **c)** the amnion.   **d)** the placenta.

_____ 5. The menstrual cycle is a series of changes in a female reproductive system that occur about
   a) once a week.   **b)** once a month.   **c)** once a year.   **d)** twice a year.

_____ 6. The new cell produced by fertilization is called
   a) an egg.   **b)** an amnion.   **c)** a zygote.   **d)** an oviduct.

_____ 7. The ovaries produce both hormones and
   a) sperm.   **b)** eggs.   **c)** zygotes.   **d)** urine.

_____ 8. An embryo receives nourishment and gets rid of wastes through a thick, flat structure called the
   a) amnion.   **b)** umbilical cord.   **c)** placenta.   **d)** uterus.

_____ 9. A zygote attaches to the wall of the
   a) uterus.   **b)** ovary.   **c)** vagina.   **d)** cervix.

_____ 10. The process by which blood and the tissue lining leave the uterus is
   a) ovulation.   **b)** birth.   **c)** menstruation.   **d)** meiosis.

_____ 11. The main organs of the male reproductive system are
   a) sperm cells.   **b)** urethras.   **c)** testosterones.   **d)** testes.

_____ 12. When a mature egg is released from the ovary, it passes through the
   a) vagina.   **b)** cervix.   **c)** uterus.   **d)** oviduct.

_____ 13. Which of the following only can occur during ovulation?
   a) menstruation   **b)** fertilization   **c)** meiosis   **d)** mitosis

WORD SCRAMBLE

Below are several scrambled words you have used in this Lesson. Unscramble the words and write your answers in the spaces provided.

1. YOBREM _____

2. YZTOEG _____

3. TACNELAP _____

4. TUSEF _____

5. ZATFERNIOTLII _____

REACHING OUT

At birth, the placenta is still connected to the infant by the umbilical cord. Why do doctors remove the umbilical cord upon birth?

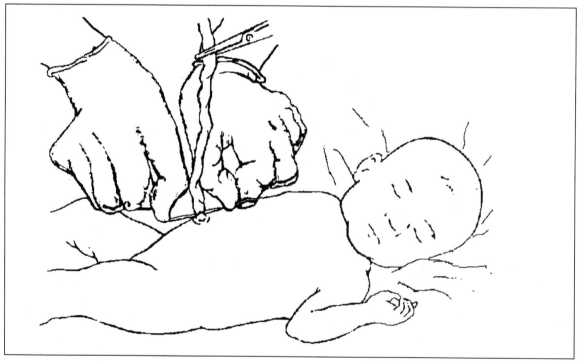

Figure K

_____

_____

_____

What are the stages of human development?

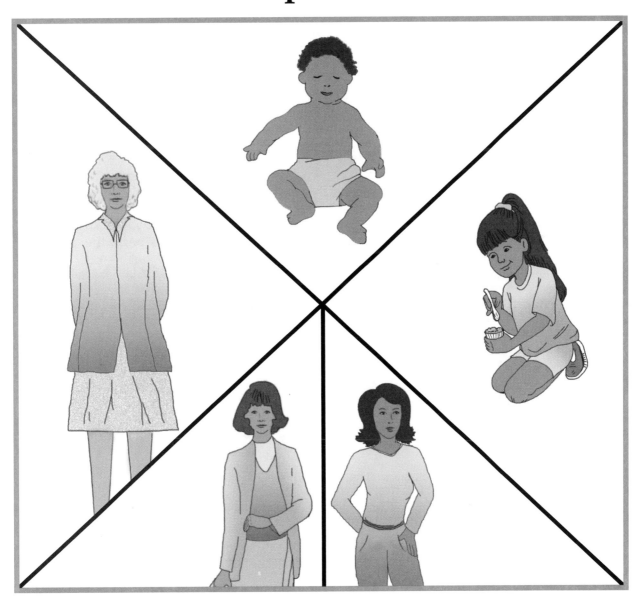

LESSON 28 | What are the stages of human development?

As you have learned in the previous lesson, a developing baby goes through a series of stages of development. Once the baby is born, the development continues. Human development begins at birth and continues into old age.

At birth, all of the major body systems and organs are in place. However, it often takes years before all of the body systems are fully developed and capable of working properly.

The series of stages a person goes through is called a life cycle. The human life cycle has five stages. These stages are: infancy, childhood, **adolescence** [ad-ul-ES-uns], adulthood, and old age.

At each stage of the human life cycle, different events happen, making that stage unique.

Infancy is the earliest stage of human development. Infancy begins at birth and ends at age 2. At birth, infants are totally helpless. They depend on others for everything.

Infancy is marked by rapid growth. The muscular system and nervous system develop quickly. Mental skills develop as the infant interacts with its surroundings. By the end of the infancy stage, most infants can walk and are able to speak.

Figure A

CHILDHOOD

Childhood is generally defined as the period between the ages 2 and 10 or 12. During childhood more complex development of the muscular and nervous systems occurs. Children also grow taller and gain weight. Their first set of teeth are replaced by permanent teeth.

During this period, children become more self-reliant. They do not need to rely on others for as many tasks as they did as infants. Children are able to feed and dress themselves. Mental abilities expand greatly. Most children start to learn to read and write during childhood.

Figure B

Adolescence begins between the ages of 10 and 12. During this stage, the young person undergoes very rapid growth "spurts." The body gets taller and stronger.

Other physical changes also take place. The onset of adolescence is called puberty. The sex organs mature. Secondary sex characteristics develop. Adolescents of both sexes are now capable of reproduction.

Figure C

Puberty and Body Changes

The two reproductive organs most affected by puberty are the ovaries and the testes. During adolescence, the ovaries and testes mature. The ovaries produce eggs. The testes produce sperm. However, the ovaries and testes have another job as well. They produce hormones. These hormones are carried to all parts of the body by the blood. These hormones help in the development of secondary sex characteristics.

The female sex hormone produced by the ovaries is **estrogen** [ES-truh-juhn]. Estrogen is responsible for the development of breasts, growth of body hair, the widening of the hips, and the beginning of menstruation.

The male sex hormone produced by the testes is **testosterone** [tes-TAWS-tuhr-ohn]. Testosterone is responsible for the production of sperm and semen, growth of body hair, growth in size of the penis and testes, increase in shoulder and muscle size, and the deepening of the voice.

For both sexes, after puberty, general growth slows down.

Figure D

Adulthood begins usually between the ages of 18 and 21. It is marked by the completion of physical growth. Muscle development and coordination reaches its peak during adulthood.

OLD AGE

Figure E

Old age is the beginning of the aging process. Between the ages of 30 and 50, muscle tone begins to decrease. Other signs of aging are a decrease in physical strength and coordination. Sense organs, such as the eyes and ears may not function as well as they once did. The bones may become brittle and may break more easily.

Aging occurs at different times in different people. The onset of aging depends upon the attitudes and habits of the individual. People who have eaten properly and exercised regularly may not show signs of aging until their 70s. People who smoke, drink alcohol, or use drugs will age faster than people who do not.

For this reason, it is important to think about the future today. The habits you develop now greatly affect aging.

Answer each question in the space provided.

1. What are some signs of aging? _____

   _____

2. What are some ways the aging process can be slowed? _____

   _____

3. How would you compare the changes that take place during infancy with those of

   old age? _____

   _____

   _____

4. How do the bones change during old age? _____

   _____

5. Why do males and females go through a period of rapid physical change between

   the ages of 11 and 14? _____

   _____

COMPLETE THE CHART

Place the letter of each statement under the stage of the human life cycle that it describes. You will use some letters more than once.

Infancy	Childhood	Adolescence	Adulthood	Old Age

a. Some organs are not fully developed.

b. Physical growth is completed.

c. Puberty takes place.

d. Vocabulary develops.

e. Muscle development is completed.

f. Ability to reproduce is present.

g. Sense receptors get weaker.

h. Powers of reasoning develop.

i. Muscle and nerve development occur rapidly.

What is meant by good health?

LESSON 29 | What is meant by good health?

"How are you feeling?" . . . "OK!" . . . This is the most commonly asked question, and reply. But what does "OK" mean? To some people, "OK" means "not being sick" and nothing more. However, good health means more than "not being sick." It means feeling happy and full of energy. It means accepting and adjusting to everyday stresses, disappointments, and challenges. In short, good health means "not just living" but ENJOYING life as well.

There are four basic factors that are needed for good health. They include proper rest, regularly exercising, eating a balanced diet, and maintaining your weight.

BALANCED DIET   You have already learned about eating a balanced diet. A balanced diet provides you with the energy and nutrients you need for daily activities, growth, and body maintenance.

MAINTAIN YOUR WEIGHT   The amount of food you eat each day and the amount of energy your body needs plays an important part in maintaining your weight. If you eat more food than your body needs, you will gain weight. If you eat less food than what your body needs, you will lose weight. During adolescence, a lot of energy is put into growing, so it is important to feed your body the energy it needs.

EXERCISE REGULARLY   Exercise also is related to maintaining weight. By exercising, you use more energy. This often leads to weight loss. But exercise also is important for good health. Regular exercise strengthens your heart and other muscles. It promotes better posture and improves endurance. When you are "fit" you often have a better self-image.

PROPER REST   Rest is just as important as the other elements of good health. Sleep is the best form of rest. Every person needs a different amount of rest each day. If you get too little rest, your body will weaken and be prone to illness.

Good health involves not only physical well-being, but mental and social well-being as well. Decisions that you make today will affect your health in the future.

What are some of the benefits of good health? People that are healthy are more energetic. Their body systems work better. Healthy people have more control over their emotions and stress. Healthy people have more confidence. They develop a better self-image. Truly good health does not happen by itself. You have to work at it. The results are worth it!

Good health involves more than exercise, diet, and rest. Other things are important too.

Figure A *Reduce the amount of salt and fat in your diet.*

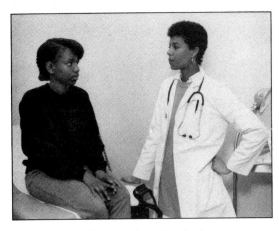

Figure B *Get regular physical examinations.*

Figure C *Brush and floss teeth regularly. Get regular dental checkups.*

Figure D *Always wear your seat belt in the car and follow other safety rules while driving.*

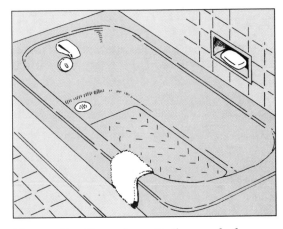

Figure E *Keep clean. Bathe regularly.*

Figure F *Do not smoke, drink, or take drugs.*

FILL IN THE BLANK

Complete each statement using a term or terms from the list below. Write your answers in the spaces provided. Some words may be used more than once.

adjusts	exercise	sleep
alcohol	rest	tobacco
balanced diet	salt	weight
drugs	sick	

1. If you eat more food than your body needs for energy you will gain

   _____ .

2. The main source of rest is _____ .

3. _____ strengthens your heart and other muscles.

4. The four main components of good health are: maintain a proper _____ ,

   _____ regularly, eat a _____ , and get plenty of

   _____ .

5. Being healthy is more than just not being _____ .

6. A truly healthy person _____ to everyday problems.

7. Being healthy also includes "saying no" to _____ , _____ ,

   and _____ .

8. You should try to limit your fat and _____ intake to maintain good

   health.

REACHING OUT

There are many other aspects of good health not mentioned in this lesson. See how many others you can think of. Write your answers in the space provided. Then share your ideas with those of your classmates.

_____

_____

_____

How do drugs affect the body?

KEY TERMS

drug: any chemical substance that causes a change in the body

drug abuse: improper use of a drug

LESSON 30 | How do drugs affect the body?

Have you ever taken drugs? Think first before answering. A **drug** is any chemical substance that causes a change in the body. Drugs can cause physical and/or mental changes in the body. They can also change behavior.

Drugs can be helpful. They can cure and prevent disease. They can relieve pain. There are two kinds of drugs for medical use, over-the-counter drugs and prescription drugs. Aspirin, decongestants, and antacids are common over-the-counter drugs. You can buy them in any store without a doctor's permission. Prescription drugs, however, can be purchased only with a written prescription from a doctor. All drugs, prescription and over-the-counter, should be taken with great care. You should always follow label directions.

Most people use drugs wisely. Some people, unfortunately abuse them. **Drug abuse** is the improper use of a drug. How are drugs abused? Sometimes people take too much of a drug, or take it for the wrong reason. The use of illegal drugs also is drug abuse. Why do you think some people abuse drugs?

Many drug abusers become physically or emotionally dependent on a drug. This means the body cannot get along without the drug. Other problems of drug abuse include **tolerance** [TOL-uhr-uns]. Tolerance occurs when the body gets used to the drug. The individual needs stronger and stronger doses to achieve the same effect. Tolerance can lead to a drug overdose and even death.

In order to treat drug abuse, a person must first go through withdrawal. Symptoms of withdrawal include chills, fever, vomiting, and even convulsions. These symptoms can last as short as a few days or as long as a few weeks.

There is a much better way to avoid all of these symptoms.

SAY NO TO DRUGS from the very beginning.

Scientists classify drugs according to the effects they have on the body.

Figure A

Stimulants Drugs that speed up the action of the central nervous system are called stimulants [STIM-yuh-lents]. Stimulants speed up a person's heart rate and breathing rate. Some examples of stimulants are caffeine, cocaine, and **nicotine** [NIK-uh-teen].

Figure B

Cocaine is a commonly abused stimulant. Crack is a purified form of cocaine. Caffeine also is commonly abused. Caffeine is found in coffee, tea, cola, and chocolate. Nicotine is found in tobacco.

Figure C

Depressants Depressants [di-PRESS-unts] are drugs that slow down the action of the central nervous system. They slow down a person's heart rate and breathing rate. Large amounts of depressants can cause a person to go into a coma, or even die. Some examples of depressants are alcohol, **barbiturates** [bar-BICH-uhr-its] and tranquilizers.

Barbiturates often are used in sleeping pills. They also have many other medical uses.

Figure D

Narcotics Narcotics [nar-KOT-iks] are depressant drugs made from the opium plant. Some common examples of narcotics include morphine and codeine. Morphine and codeine are often used as pain killers. Another narcotic is heroin. Heroin is an illegal narcotic that has no medical use.

Figure E

Hallucinogens Hallucinogens [huh-LOO-suh-nuh-jenz] are drugs that distort or alter the senses. LSD and marijuana are two commonly abused hallucinogens. Hallucinogens cause a person to feel panic or threatened. For this reason, people who take hallucinogens are dangerous to themselves and to others.

Hallucinogens also cause the brain to "see" things that are not really there—figures, designs, color, movement. For example, an LSD user may feel capable of doing impossible things, such as flying without a plane.

Figure F

Marijuana is the most widely abused illegal drug in the United States. It comes from a plant and usually is smoked. Marijuana causes the user to have hallucinations, but also slows down the central nervous system. For this reason, marijuana often is classified as a depressant.

COMPLETE THE CHART

Decide whether each drug listed in the chart is a **stimulant, hallucinogen, narcotic,** or **depressant**. Complete the chart by writing the name of the group to which the drug belongs in the column at the right.

	Drug	Classified As
1.	Caffeine	
2.	Barbiturate	
3.	Nicotine	
4.	Crack	
5.	Marijuana	
6.	Cocaine	
7.	LSD	
8.	Alcohol	
9.	Morphine	
10.	Tranquilizer	
11.	Heroin	
12.	Codeine	

MATCHING

Match each term in Column A with its description in Column B. Write the correct letter in the space provided.

Column A

_____ 1. drug

_____ 2. addiction

_____ 3. depressant

_____ 4. drug abuse

_____ 5. tolerance

Column B

a) any chemical that causes a change in the body

b) drug that slows down the central nervous system

c) uncontrollable dependence on a drug

d) when the body gets used to a drug

e) improper use of a drug

READING DRUG LABELS

What You Need (Materials)

3 different over-the-counter drug or medication labels or packaging

How To Do The Experiment (Procedure)

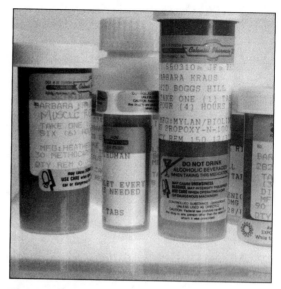

Figure G

1. Ask your parents for labels from over-the-counter medicines. Examine each one of your drug or medication labels.

2. Write the name of the drug or medication and its use.

3. Find the dosage instructions and record the dosage.

4. Find the expiration date and record it.

5. Record any special warnings or precautions given for the drug or medication.

What You Learned (Observations)

1. What two things are given in the dosage instructions for a drug or medication?

   _____

   _____

2. Do the dosages differ for different age groups? If so, give an example. _____

   _____

3. What frequencies for taking the drugs or medications did you find? _____

   _____

Something To Think About (Conclusions)

1. List the types of information that you can obtain from drug or medication labels.

   _____

   _____

   _____

   _____

2. What general warning statement is found on all drug labels?

   _____

How does alcohol affect the body?

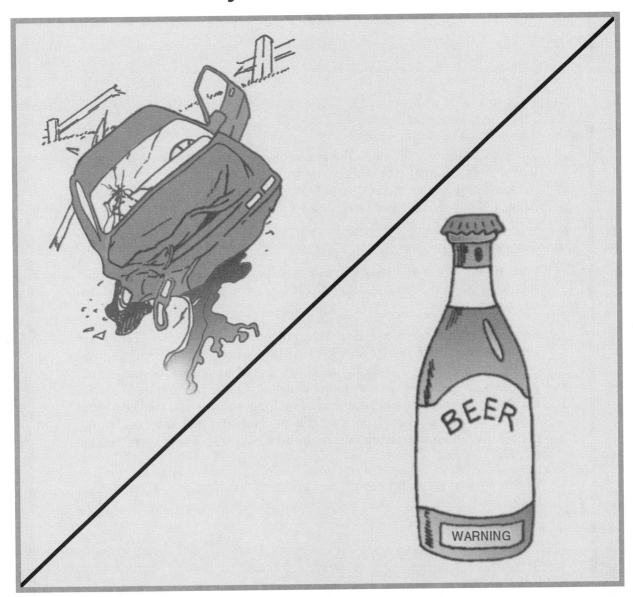

KEY TERMS

alcoholic: person who is dependent on alcohol

cirrhosis: liver disorder caused by damaged liver cells

LESSON 31 | How does alcohol affect the body?

The use, or more accurately, the abuse of alcohol is a growing problem. It demands national attention. What is alcohol and how does it affect a person? Simply and bluntly stated, alcohol is a <u>drug</u>, a nonprescription drug. It's a drug that the law allows you to buy and to use, if you are over the "legal drinking age."

Alcohol is all around us. How many liquor stores and bars are there in your town? Open any magazine or newspaper, and you will probably see many ads for beer and other alcoholic beverages. The same goes for television.

People drink for many reasons. They drink at meals, to be sociable, for celebrations. Other people drink to be part of the "in crowd." Others drink to ease stress. In fact, name any event or situation, and some people will use that as a reason to drink. Often one drink leads to another, and another, and another. Before long, some individuals do little more than drink. They have become dependent upon alcohol. These people have become **alcoholics** [al-kuh-HOWL-iks]. Their body craves alcohol.

In this lesson, we will examine the effect alcohol has on the mind and body.

Many people feel that alcohol is a stimulant drug. It may be, but just for a short time. Then the alcohol slows down the body and mind. In reality, alcohol is a depressant, a "real downer." Alcohol abuse can cause many health and personal problems.

BLOOD ALCOHOL CONCENTRATION (BAC)

The amount of alcohol present in the bloodstream is called the Blood Alcohol Concentration (BAC). The effect of alcohol on the body increases as the level of alcohol in the bloodstream increases. The way alcohol affects someone depends upon many different factors. Figure A shows the Blood Alcohol Concentration (BAC) level and its effects on the body.

Drinks Per Hour	BAC (percent)	Effects
1	0.02–0.03	Feeling of relaxation
2	0.05–0.06	Slight loss of coordination
3	0.08–0.09	Loss of coordination, slurred speech and trouble thinking
4	0.11–0.12	Lack of judgment, increased reaction time
7	0.20	Difficulty thinking, walking, talking
14	0.40	Unconsciousness, vomiting
17	0.50	Deep coma; if breathing stops, death occurs
One drink = 1 oz. whiskey, 4 oz. wine, or 12 oz. beer		

Figure A *Blood Alcohol Concentration and Its Effects*

1. What does BAC stand for? _____

2. At what BAC level does unconsciousness or vomiting occur? _____

3. What is the effect of drinking one drink? _____

4. After how many drinks does slurred speech and loss of coordination occur? _____

5. How much whiskey is equivalent to 12 oz. of beer? _____

6. What can result from BAC of .50? _____

7. If one drink is made with 2 ounces of whiskey, what BAC level will result? _____

8. Why would it be dangerous to drive after having 4 drinks? _____

9. At what BAC level does a person experience difficulty thinking, walking, and

talking? _____

10. What is alcoholism? _____

EFFECT OF ALCOHOL ON THE BODY

Circulation   Alcohol damages the heart as well as the blood vessels. Alcoholics often have heart disease as well as high blood pressure. Excessive alcohol can slow down the heart rate so much, that it stops.

Digestion   Alcohol damages the lining of the esophagus, stomach and small intestine. It can cause painful ulcers. Continued drinking makes the ulcers, and the pain, worse.

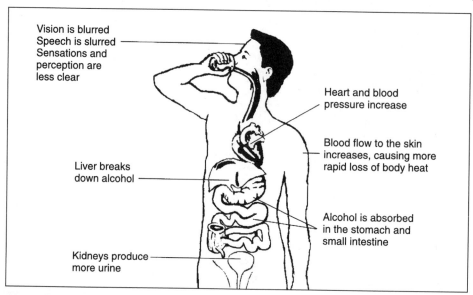

Vision is blurred
Speech is slurred
Sensations and perception are less clear

Heart and blood pressure increase

Blood flow to the skin increases, causing more rapid loss of body heat

Liver breaks down alcohol

Alcohol is absorbed in the stomach and small intestine

Kidneys produce more urine

Figure B

Heavy drinking also damages the liver. Alcohol breaks down the liver cells. As the liver cells are damaged or destroyed, they are replaced with scar tissue. This condition is called **cirrhosis** [suh-ROH-sis] of the liver. Eventually the liver stops functioning. Cirrhosis of the liver is a major cause of death among heavy drinkers. Figure C shows a liver damaged by cirrhosis.

Figure C   *A healthy liver, a fatty liver, and a liver damaged by cirrhosis*

What effect does alcohol have on the liver? _____

_____

Further liver damage is caused by malnutrition. Drinkers usually have poor diets, and therefore do not receive all the proper minerals and vitamins.

EFFECT OF ALCOHOL ON THE MIND

Alcohol affects people in different ways. Alcohol can change a person's personality. A normally peaceful person may become aggressive. This can lead to violent behavior.

Heavy drinking can cause a person to lose his or her job. It can lead to divorce and family break-up. Some alcoholics end up as homeless people who must beg for money.

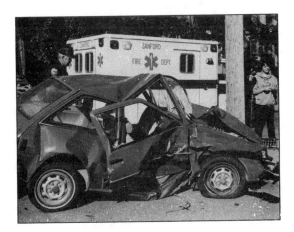

Figure D

Alcohol use and abuse can kill, not only the drinker, but innocent people too. Alcohol blurs vision, affects judgment and slows reaction time. Tragically, auto accidents caused by "alcohol impaired" drivers are on the rise. About one-half of all traffic deaths are alcohol related.

HELP FOR ALCOHOLICS

Alcoholism is a disease. However, there is a help for alcoholics and their families. Many alcoholics get help from groups such as Alcoholics Anonymous. Teenage children of alcoholic parents or siblings can get help from groups such as Alateen.

Imagine that you have a relative or friend who has a drinking problem. What would you

do to help this person? _____

_____

REACHING OUT

Why do you think it is dangerous to mix alcohol and drugs (even "harmless" drugs such as aspirin)?

_____

_____

SCIENCE *EXTRA*

Radiology Technician

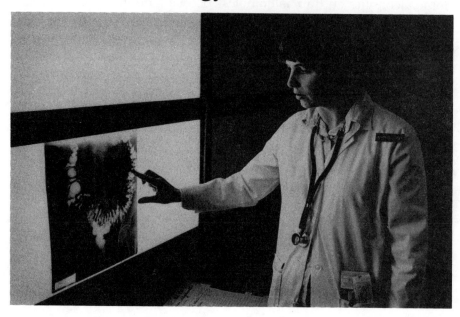

Did you ever have an X-ray taken? Almost everyone has—either by a dentist or a doctor. X-rays are forms of radiation. They are important in many fields. They are especially important for medical and dental diagnosis. They enable physicians and dentists to see outlines of internal organs. Ordinarily, these parts cannot be seen from the outside. Taking an X-ray allows medical people to diagnose problems without cutting into the body.

X-rays not only help diagnose many ailments. They also treat certain diseases—such as cancer. First, the X-rays help to diagnose the cancer. Then, concentrated and carefully aimed X-rays (or other radioactive sources) destroy the cancer cells. If the cancer is diagnosed early and has not spread, the cure may be total.

X-rays can help people, but radiation also can harm living things. Because radiation can harm living things, it is important to check the amount of radiation used in the X-ray procedure. It is the job of an X-ray technician to operate X-ray equipment and prepare patients for X-rays.

X-ray technicians fill a very important role in medical diagnosis and treatment. Their special skills are always in demand. They are employed by hospitals, clinics, and group medical organizations. Radiology technicians are licensed by individual states. But most standards are set by the federal government.

If you do well in science and math, you might consider training for this steady, and well-paying profession. To become an X-ray technician, you must first graduate from high school. An X-ray technician also must complete a two-year X-ray technician program.

How does tobacco affect the body?

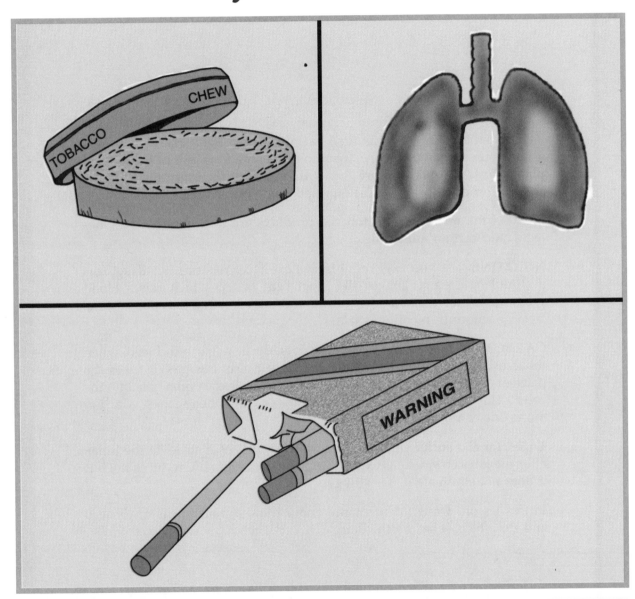

KEY TERMS

tar: sticky, yellow substance found in tobacco

nicotine: stimulant drug found in tobacco

carbon monoxide: poisonous gas produced when tobacco is burned

LESSON 32 | How does tobacco affect the body?

What is tobacco and why is it so harmful to the body? Have you ever asked yourself that question? Tobacco is the shredded leaf of the tobacco plant. Tobacco contains more than a thousand different products. Many of these products are harmful especially when they are smoked.

Three of the most harmful tobacco products are **tar**, **nicotine** [NIK-uh-teen], and **carbon monoxide**.

NICOTINE As you may recall from Lesson 31, nicotine is a drug. It is a stimulant drug. It makes the heart beat faster, which raises blood pressure. It increases blood pressure and damages the nervous system. In large amounts, nicotine is fatal.

CARBON MONOXIDE Carbon monoxide is a gas produced when tobacco burns. Carbon monoxide is a very poisonous gas. It takes the place of oxygen in the blood stream. As a result, less oxygen gets into the cells. This causes dizziness, drowsiness, and headaches. Carbon monoxide can damage the brain.

TAR Tar is a sticky yellowish substance. Much of it sticks to the lungs after the tobacco smoke is exhaled. A filter on a cigarette reduces tar, but it does not eliminate it. Tar still gets into the lungs.

In this lesson, we will learn more about the different forms of tobacco and the effects it has on the body.

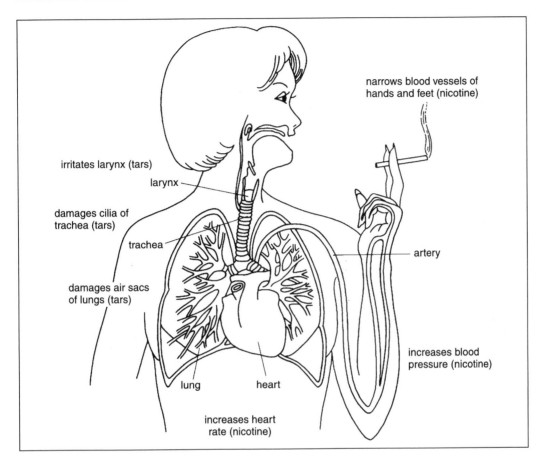

Figure A

Use Figure A and what you have already learned to answer the following questions.

1. What is tobacco? _____

   _____

2. What are the three harmful products found in tobacco?

   _____   _____   _____

3. What effect does carbon monoxide have on the body? _____

   _____

4. What effect does nicotine have on the body? _____

   _____

5. What happens to tar after a smoker breathes out tobacco smoke? _____

   _____

Tobacco products have been linked to many kinds of cancer. Smoking is the major cause of two deadly lung diseases, lung cancer and **emphysema** [em-fuh-SEE-muh]. A person with emphysema often has shortness of breath.

Smoking cigarettes is not the only culprit. People who smoke pipes or cigars also develop cancers. Smokers often develop cancer of the lungs, bladder, kidneys, pancreas, mouth, larynx, esophagus, cheek, lip, and tongue.

Chewing tobacco causes cancer of the mouth. In fact, cancer from chewing tobacco often develops faster than other kinds of cancer.

YOU BE THE DOCTOR

The symptoms of emphysema and lung cancer are described below. Read the description and see if you can identify the illness.

Illness A   Tobacco smoke damages the lungs' air sacs. They become clogged with phlegm. Fresh air cannot enter. Stale air becomes trapped and cannot be exhaled. The walls of the air sacs break down. This leaves large spaces inside the lungs. Breathing becomes difficult. Oxygen supply to the body is reduced. The heart tries to make up for this by working harder and harder. Finally the heart stops. The result is death from heart failure.

Illness B   Normal cells become abnormal. They multiply rapidly. As they multiply, they push aside and destroy healthy cells. The abnormal cells usually spread to other parts of the body. The life processes are reduced so much that death occurs.

1. Which illness is cancer? _____

2. Which illness is emphysema? _____

Ask yourself this question, "Is it worth it?" If you are a smoker, stop NOW, before it is too late. If you do not smoke, NEVER start.

FILL IN THE BLANK

Complete each statement using a term or terms from the list below. Write your answers in the spaces provided. Some words may be used more than once.

nicotine   carbon monoxide   tar
pipes   plant   heart
cigars   cancer   chewing tobacco
stimulant   emphysema

1. What are three harmful substances in tobacco? _____

   _____ _____

2. Nicotine is a _____ drug.

3. A sticky yellowish substance is _____ .

4. The poisonous gas caused by burning tobacco is _____ .

5. Tobacco used in cigarettes comes from the tobacco _____ .

6. Other forms of smoking include the use of _____ and

   _____ .

7. Cancer of the mouth often is related to use of _____ .

8. Shortness of breath is a symptom of _____ .

9. _____ is the rapid division of abnormal cells.

10. Smoking puts a great strain on the _____ .

MATCHING

Match each term in Column A with its description in Column B. Write the correct letter in the space provided.

	Column A		Column B
_____	1. carbon monoxide	a)	yellow sticky substance found in tobacco
_____	2. tar	b)	characterized by shortness of breath
_____	3. nicotine	c)	abnormal cells replacing healthy cells
_____	4. emphysema	d)	poisonous gas
_____	5. lung cancer	e)	stimulant drug

You do not have to be a smoker to suffer from smoke-related problems. If you are in the same room with a smoker, you will breathe in their smoke. Evidence indicates that this smoke, often called secondhand smoke, is just as damaging as the smoke from active smoking.

Figure B

Many organizations have now banned smoking or restricted smoking areas. This has caused arguments between smokers and nonsmokers. The smokers claim they have the right to smoke anywhere. The nonsmokers claim that they should not be exposed to harmful smoke.

Who do you think is right? In the space below, write your argument for or against the smokers. There is no right or wrong answer, but you should be able to prove your position.

_____

_____

_____

_____

_____

CLOTHING PROTECTION • A lab coat protects clothing from stains. • Always confine loose clothing.

 EYE SAFETY • Always wear safety goggles. • If anything gets in your eyes, flush them with plenty of water. • Be sure you know how to use the emergency wash system in the laboratory.

 FIRE SAFETY • Never get closer to an open flame than is necessary. • Never reach across an open flame. • Confine loose clothing. • Tie back loose hair. • Know the location of the fire-extinguisher and fire blanket. • Turn off gas valves when not in use. • Use proper procedures when lighting any burner.

 POISON • Never touch, taste, or smell any unknown substance. Wait for your teacher's instruction.

 CAUSTIC SUBSTANCES • Some chemicals can irritate and burn the skin. If a chemical spills on your skin, flush it with plenty of water. Notify your teacher without delay.

 HEATING SAFETY • Handle hot objects with tongs or insulated gloves. • Put hot objects on a special lab surface or on a heat-resistant pad; never directly on a desk or table top.

 SHARP OBJECTS • Handle sharp objects carefully. • Never point a sharp object at yourself—or anyone else. • Cut in the direction away from your body.

 TOXIC VAPORS • Some vapors (gases) can injure the skin, eyes, and lungs. Never inhale vapors directly. • Use your hand to "wave" a small amount of vapor towards your nose.

 GLASSWARE SAFETY • Never use broken or chipped glassware. • Never pick up broken glass with your bare hands.

 CLEAN UP • Wash your hands thoroughly after any laboratory activity.

 ELECTRICAL SAFETY • Never use an electrical appliance near water or on a wet surface. • Do not use wires if the wire covering seems worn. • Never handle electrical equipment with wet hands.

 DISPOSAL • Discard all materials properly according to your teacher's directions.

GLOSSARY/INDEX

absorption [ab-SAWRP-shun]: movement of food from the digestive system to the blood, 66

alcoholic [al-kuh-HOWL-ik]: person who is dependent on alcohol, 188

alveoli [al-VEE-uh-ly]: microscopic air sacs in the lungs, 98

amino acid: building block of proteins, 35

arteries [ART-ur-ees]: blood vessels that carry blood away from the heart, 70

atria [AY-tree-uh]: upper chambers of the heart, 82

behavior: response of an organism to its environment, 144

bile: green liquid that breaks down fats and oils, 60

bronchi [BRAHN-kee]: tubes leading to the lungs, 96

capillaries [KAP-uh-ler-ees]: tiny blood vessels that connect arteries to veins, 70

carbohydrate [kar-buh-HY-drayt]: nutrient that supplies energy, 32

carbon monoxide: poisonous gas produced when tobacco is burned, 194

cardiac [KAHR-dee-ak]: type of muscle found only in the heart, 20

cartilage [KART-ul-idj]: tough, flexible connective tissue, 14

cerebellum [ser-uh-BELL-um]: part of the brain that controls balance and body motion, 126

cerebrum [suh-REE-brum]: large part of the brain that controls the senses and thinking, 126

cirrhosis [suh-ROH-sis]: liver disorder caused by damaged liver cells, 190

conditioned [kun-DISH-und] **response**: behavior where one stimulus takes the place of another, 144

digestion [dy-JES-chun]: process by which foods are changed into forms the body can use, 52

drug: any chemical substance that causes a change in the body, 182

drug abuse: improper use of a drug, 182

egg: female reproductive cell, 156

elimination [ee-LIM-uh-NA-shun]: the removal from the body of wastes from digestion, 101

embryo [EM-bree-oh]: hollow ball of cells formed by cell division of the zygote, 166

endocrine [EN-duh-krun] **system**: body system made up of about ten endocrine glands that help the body respond to changes in the environment, 138

enzyme [EN-zym]: protein that controls chemical activities, 58

esophagus [i-SAF-uh-gus]: tube that connects the mouth to the stomach, 52

excretion [ik-SKREE-shun]: the removal from the body of wastes made by the cells, 102

excretory system: body system responsible for removing cellular wastes from the body, 106

fat: energy-storage nutrient, 32

fertilization [fur-tul-i-ZAY-shun]: joining of one sperm cell and one egg cell, 164

habit: learned behavior that has become automatic, 150

hormone [HAWR-mohn]: chemical messengers that regulate body functions, 138

joint: place where two or more bones meet, 14

ligaments [LIG-uh-ments]: tissue that connects bone to bone, 14

marrow: soft tissue in a bone that makes blood cells, 14

medulla [muh-DULL-uh]: part of the brain that controls many vital functions, such as heartbeat and breathing, 126

mineral: nutrient needed by the body to develop properly, 40

nervous system: body system made up of the brain, the spinal cord, and all the nerves that control body activities, 120

201

neuron [NOOR-ahn]:   nerve cell, 121

nicotine [NIK-uh-teen]:   stimulant drug found in tobacco, 194

nutrient [NOO-tree-unt]:   chemical substance in food needed by the body for growth, energy, and life processes, 26

organ [OWR-gun]:   groups of tissues that join together to do a specific job, 2

organ system: group of organs that work together, 8

ovaries [OH-vuhr-eez]:   female reproductive organs, 157

peristalsis [per-uh-STAWL-sis]:   wavelike movement that moves food through the digestive tract, 55

placenta [pluh-SEN-tuh]:   organ through which an embryo receives nourishment and gets rid of wastes, 166

plasma [PLAZ-muh]:   liquid part of blood, 76

platelets [PLAYT-lits]: tiny colorless pieces of cells that help blood clot, 76

protein [PRO-teen]:   nutrient needed by the body to build and repair cells, 32

red blood cells: cells that give blood its red color and carry oxygen, 76

reflex: automatic response to a stimulus, 132

respiration [res-puh-RAY-shun]:   process of carrying oxygen to cells, getting rid of carbon dioxide, and releasing energy, 90

septum: thick tissue wall that separates the left and right sides of the heart, 82

skeletal muscle: muscle attached to the skeleton, making movement possible, 20

smooth muscle: muscle that cause movements that you cannot control, 20

specialized [SPESH-uh-lyzed] **cells**:   cells that are similar in size and shape, 2

sperm: male reproductive cell, 156

tar: sticky, yellow substance found in tobacco, 194

testes [TES-teez]:   male reproductive organs, 160

tissues: group of similar cells that work together to perform a specific function, 2

trachea [TRAY-kee-uh]:   windpipe, 96

valve: thin flap of tissue that acts like a one-way door, 82

veins [VANES]:   blood vessels that carry blood to the heart, 70

ventricles [VEN-tri-culz]:   lower chambers of the heart, 82

villi [VIL-y]:   fingerlike projections on the lining of the small intestine, 66

vitamin [VYT-uh-min]:   nutrient found naturally in many foods, 40

white blood cells: cells that protect the body from disease, 76

zygote [ZY-gote]:   fertilized egg produced by fertilization, 164